LIVESTOCK, ETHICS AND QUALITY OF LIFE

The Lamb

Little lamb, who made thee?
Dost thou know who made thee
Gave thee life and bid thee feed
By the stream and o'er the mead –
Gave thee clothing of delight,
Softest clothing, woolly, bright,
Gave thee such a tender voice,
Making all the vales rejoice?
Little lamb, who made thee,
Dost thou know who made thee?

Little lamb, I'll tell thee,
Little lamb, I'll tell thee!
He is callèd by thy name,
For he calls himself a Lamb;
He is meek and he is mild,
He became a little child:
I a child, and thou a lamb,
We are callèd by his name.
Little lamb, God bless thee,
Little lamb, God bless thee!

William Blake (1757–1827)
Songs of Innocence

Contents

Biographies of Editors and Contributors

Dr Donald Bruce, *The Society, Religion and Technology Project, Church of Scotland, John Knox House, 45 High Street, Edinburgh EH1 1SR, UK.*
Donald Bruce trained in chemistry, and spent 15 years in research in nuclear energy at Sellafield and Harwell in the UK and in nuclear safety and risk assessment for the UK Nuclear Installations Inspectorate. He then studied for a Diploma in Theology at Oxford and in 1992 became Director of the Society, Religion and Technology Project of the Church of Scotland, in Edinburgh. This project was set up in 1970 to look at ethical issues arising from current and future technology.

He has specialized in biotechnology, energy, environmental and risk issues, and has become internationally recognized as a writer, speaker and broadcaster, especially on cloning issues, with which he is uniquely involved. He is a member of several European church committees on bioethics and the environment, and represented the UK at a global summit meeting of national bioethics committees and was an observer to the International Committee of Bioethics of UNESCO.

Ann Bruce, *The Society, Religion and Technology Project, Church of Scotland, John Knox House, 45 High Street, Edinburgh EH1 1SR, UK.*
Ann Bruce took a BSc in Agriculture at the University of London, Wye College, and an MSc in Animal Breeding at Edinburgh University. She worked for over a decade in the animal breeding industry in the UK and several other countries and continents being concerned with quantitative genetics.

From 1993 to 1998 she and her husband, Donald, were respectively Rapporteur and Chair of an expert multi-disciplinary working group on the ethical and social aspects of genetic engineering in non-human species. The members of this study including leading figures in Scotland in animal genetics, welfare and cloning. They co-edited the book of this study, *Engineering Genesis*, published by Earthscan in November 1998.

Dr Gary L. Comstock, *Bioethics Program, Iowa State University, Ames, IA 50011-1306, USA.*
Dr Gary L. Comstock is Coordinator of the Bioethics Program at Iowa State University, USA where he is Associate Professor of Religious Studies in the Philosophy Department. He has served as President of the Society for Agriculture, Food and Human Values. He has written extensively on ethical issues associated with animal biotechnology. He is editor of the book *Is There a Moral Obligation to Save the Family Farm?*, and has been guest editor of three issues of the *Journal of Agricultural and Environmental Ethics*. He is a well-known speaker on issues related to ethics and agriculture.

Dr E. David Cook, *Whitefield Institute and Green College, Oxford University, Oxford OX1 3HZ, UK.*
Dr David Cook is a Scot with degrees in philosophy and theology from Arizona State, Edinburgh and Oxford Universities. He was a US government consultant with the Black Power Movement. His academic career includes appointments at St John's Theological College, UK, and Nottingham University. Since 1979 he has been at Oxford University, first at Westminster College and currently Fellow and Chaplain of Green College and Director of the Whitefield Institute for Research in Theology, Education and Ethics.

He broadcasts regularly with the BBC on ethical and theological issues, has had his own radio and television series and is a commentator on moral and theological issues, also advising the British Medical Association. He has given evidence to both the UK Houses of Parliament on medical ethics and on science and genetics. He is a member of the UK Xenotransplant Interim Regulatory Authority. He speaks internationally and is the author of a wide range of books including *The Moral Maze*, *Living in the Kingdom*, *Patient's Choice* and *Blind Alley Beliefs*.

Professor Denis Goulet, *University of Notre Dame, 0–119 Hesburgh Center, Notre Dame, IN 46556, USA.*
Professor Denis Goulet is O'Neill Professor in Education for Justice, Department of Economics, University of Notre Dame, USA. He is also a Faculty Fellow in both the Helen Kellogg Institute for International Studies, and the Joan B. Kroc Institute for International Peace Studies. A development ethicist with degrees in philosophy (US), social planning (France) and political science (Brazil), he has held Visiting Professorships in Brazil, Poland, France, Canada and the USA. He has extensive field experience in Africa, Latin America, Asia and the Middle East. Professor Goulet is a prolific author of books, articles and monographs on development ethics.

Professor In K. Han, *College of Agriculture and Life Sciences, Seoul National University, Suweon 441-744, Korea.*
Professor Han is Professor of Animal Nutrition and Feed Science in Seoul National University, Korea where he has had a long career in teaching, research and administrative responsibilities since 1965. Professor Han is Past President of the World Association for Animal Production of which he was President from 1993 to 1998.

Professor Han has served a variety of national bodies concerned with research, agricultural policy, the environment, the agricultural feed industry and international cooperation. He has been President of the Korean Societies of Animal Science, the Korean Nutrition Society, the Korean Society of Nutrition and Feed, and the Asia–Australian Association of Animal Production (AAAP). He is the author of many scientific papers and books and has been active in editing and publishing scientific journals. He holds distinguished awards from Korea and several other countries.

Professor R. Brian Heap, *St Edmund's College, Cambridge CB3 OBN, UK, and The Babraham Institute, Babraham, Cambridge CB2 4AT, UK.*
Professor Brian Heap is Master of St Edmunds College, Cambridge University, UK, a Fellow and Foreign Secretary of The Royal Society. He is an eminent animal biologist and until recently was Director of the Babraham Institute, Cambridge and Director of Science, UK Biotechnology and Biological Sciences Research Council. He has degrees in animal physiology and holds distinguished awards for his scientific work. He has authored many scientific papers and is a well-known speaker on the ethics of applying science to livestock.

Dr John Hodges, *Lofererfeld 16, A-5730 Mittersill, Austria.*
Dr John Hodges was responsible for animal breeding and genetic resources in the Food and Agriculture Organization of the UN. Previously he was Professor of Animal Genetics at the University of British Columbia, Canada, and earlier taught at Cambridge University, UK. He is former Head of the Production Division of the Milk Marketing Board of England and Wales. He has degrees in agriculture, livestock production, genetics and business administration from Reading and Cambridge Universities (UK) and Harvard University (USA).

He is experienced in livestock in all regions of the world. He is co-author of the book *Tropical Cattle: Origins, Breeds and Breeding Policies* (1997) which, in addition to reviewing the origins of cattle and current breeds, makes an analysis of genetic gain in the last 40 years and concludes that little or no genetic gains have been made in tropical cattle. The book presents a strong case for new development strategies for cattle improvement in the tropics which are better adapted to the cultural and socio-economic conditions. Dr Hodges now works privately as a consultant and author. He is an Editor of *Livestock Production Science*, the official journal of the European Association for Animal Production. He frequently teaches in the Newly Independent States (NIS) of the former USSR and in Central European countries.

Dr Mahabub Hossain, *International Rice Research Institute, PO Box 933, Manila, Philippines.*
Dr Mahabub Hossain has been Head of the Social Sciences Division of the International Rice Research Institute (IRRI), Manila, Philippines, since 1992. He began his career as a Staff Economist in the Bangladesh Institute of Development Studies in 1970 and rose through the ranks to serve as the Director General of this development policy research institute from 1989 to 1992. He also served the International Food Policy Research Institute, Washington, DC, USA, as a visiting scientist from 1985 to 1987, when he published the highly acclaimed IFPRI Research Reports on the Grameen Bank Credit Program and the Development Impact of Rural Infrastructures. At IRRI he has been leading the project on 'Socio-economic study, technology impact and policy analysis' and providing support to 'Management in planning and prioritization of rice research issues'.

Dr Hossain took his PhD in economics in 1977 from the University of Cambridge, England. He is a fellow of the Economic Development Institute of the World Bank. He was awarded the first gold medal by the Bangladesh Association of Agricultural Economists for his outstanding contribution to the understanding of the rural economy of Bangladesh.

Dr Shin-haeng Huh, *President, Korea Consumer Protection Board. Present address: Seoul Agricultural and Marine Products Corporation, 600 Garah-dong, Song pa-ku, Seoul 138–701, Korea.*
Dr Huh has been President of the Korea Consumer Protection Board since June 1995. Previously he was Professor of Kangwon University and Korean Minister of Agriculture, Forestry and Fisheries. Earlier he served successively as Senior Fellow, Vice President and President of the Korea Rural Economics Institute (KREI) from 1978 to 1993. Dr Huh gained his PhD at the University of Minnesota, USA; his MA from Western Illinois University, USA, and his BS at Seoul National University, Seoul, Korea.

Dave M. Juday, *Adjunct Fellow, Hudson Institute's Centre For Global Food Issues, PO Box 202, Churchville, VA 24421, USA.*
David Juday is an agricultural economist. He is Senior Associate of New Venture Development Corporation, an international agricultural trading company and he is also an Adjunct Fellow of the Center for Global Food Issues of the Hudson Institute, a private non-profit, non-political organiz-ation researching public policy issues. He is a former Visiting Fellow at the Alexis de Tocqueville Institution. From 1989 to 1993, Mr Juday served the

White House as Special Assistant for Economic Policy and Agriculture to the Vice President and later as Deputy Director of Domestic Policy. Mr Juday contributes regularly to financial, economic and agricultural newspapers and magazines on international trade, competitiveness and the environment.

Dr Gurdev S. Khush, *International Rice Research Institute, PO Box 933, Manila, Philippines.*
Dr G.S. Khush is the premier rice breeder of the world and for the last 32 years has led the rice breeding program of the International Rice Research Institute (IRRI) in the Philippines. More than 300 breeding lines of rice developed under his leadership have been released as varieties in different rice growing countries. Others have been used as parents in national rice breeding programmes. It is estimated that IRRI-bred varieties or their progenies are now grown on 60% of the world's rice land. Increased rice production resulting from adoption of these breeding materials feeds 1 billion rice consumers annually.

Dr Khush is a world renowned cytogeneticist and author of a widely used text on cytogenetics. He has authored two further books, edited five books, published 70 book chapters and 130 research papers in refereed journals.

For these contributions to world food security, Dr Khush has been honoured with prestigious international awards including the Japan Prize, the World Food Prize and the Rank Prize. For his scientific contributions, Dr Khush has been elected to several of the most prestigious academies such as Indian National Science Academy, The Third World Academy of Sciences, Foreign Associate US National Academy of Sciences and The Royal Society (London). He has received Honorary Doctorate degrees from five universities.

Professor George K. Kinoti, *Department of Zoology, University of Nairobi, PO Box 30197, Nairobi, Kenya. Present address: African Institute for Scientific Research and Development (AISRED), PO Box 14663, Nairobi, Kenya.*
Professor George Kinoti was until recently Professor of Zoology at the University of Nairobi, Kenya, where he was formerly Dean of Science. He was trained in East Africa and the UK.

His research and teaching are in the field of parasitology. He has served on the boards of several national research bodies. He maintains open communication both with scientific developments in the West and life in rural Africa. He is currently engaged in a project that seeks to improve livestock production by the Maasai and other people in arid and semi-arid areas of Africa. Concern about poverty and underdevelopment in Africa led to the formation of the African Institute for Scientific Research and Development (AISRED), of which he is Executive Director, to help tackle these problems in practical ways.

Dr C.U. Leon-Velarde, *Livestock Systems Specialist, ILRI-CIP, Lima, Peru. Present address: International Livestock Research Institute, PO Box 5689, Addis Ababa, Ethiopia.*
Dr Carlos Leon-Velarde was born in Iquitos, Peru. He graduated from the National Agrarian University, La Molina, Lima, Peru, in Animal Science. He obtained an MSc in Animal Production at Tropical Agronomic Research and Training Centre (CATIE) in Turrialba, Costa Rica, and a PhD in Animal Breeding at the University of Guelph in Ontario, Canada.

He has worked internationally for the Interamerican Institute for Co-operation in Agriculture (IICA), CATIE, the International Development Research Centre (IDRC) in the Dominican Republic, Costa Rica, other Central American countries and Peru. He has also been an international consultant in Guinea-Bissau.

Currently he holds a joint appointment as livestock systems specialist for the International Potato Centre (CIP) and the International Livestock Research Institute (ILRI) based in Lima, Peru. He is in charge of several research projects for the improvement of smallholder crop–livestock systems in the Andean ecoregion.

Dr Hugo Li-Pun, *International Livestock Research Institute, PO Box 5689, Addis Ababa, Ethiopia.*
Hugo Li-Pun was born in Lima, Peru. He graduated from the National Agrarian University, La Molina, Lima, Peru, in Animal Sciences. He obtained MSc and PhD degrees in Dairy Sciences and Ruminant Nutrition from the University of Wisconsin in Madison, Wisconsin in the USA.

He worked for the private sector and at the National Agrarian University in Peru. He has been working internationally for the past two decades at the Tropical Agronomic Research and Training Centre (CATIE) and the International Development Research Centre (IDRC) based in Panama, Colombia, Uruguay and Canada. While at IDRC he was in charge of a global programme on Animal Production Systems research, which developed 60 projects in 38 countries in Latin America and the Caribbean (LAC), Asia and Africa as well as organizing several research networks and initiatives to promote international collaboration in agricultural research for development.

He is now Director of the Sustainable Production Systems Programme of the International Livestock Research Institute (ILRI), which has projects in Africa, Asia and Latin America and the Caribbean and is based in Ethiopia, where he is also the Resident Director of ILRI.

Dr V.M. Mares, *Consultant, Panama. Present address: International Livestock Research Institute, PO Box 5689, Addis Ababa, Ethiopia.*
Dr Victor Mares is a Peruvian national. He obtained a BSc in Animal Science at the National Agrarian University, La Molina, Lima, Peru, an MSc in Grassland Agronomy at the University of Wales (Wales, UK) and a PhD in Crop Science at the University of Guelph (Ontario, Canada).

He has worked in the private sector in Peru and in public and international teaching and research institutions in Peru, Panama and Costa Rica. He has been an internationally recruited staff of CATIE in Central America.

Currently, he lives in Panama where he is associated with GEA Consulting, a private company. He is an international consultant and he has carried out consulting assignments in Mexico, Honduras, Nicaragua, Costa Rica, Panama, Colombia, Ecuador, Peru, Kenya and Ethiopia for IFAD, the Interamerican Development Bank, the World Bank, the Government of Ecuador, FAO, IDRC, IICA, CIP and ILRI.

Professor Bernard E. Rollin, *Colorado State University, Fort Collins, Colorado 80523-1781, USA.*
Professor Bernard E. Rollin is Professor of Philosophy and Professor of Physiology and Director of Bioethical Planning at Colorado State University. He is the author of 12 books including *Animal Rights and Human Morality, The Unheeded Cry: Animal Consciousness, Animal Pain and Science, The Frankenstein Syndrome: Ethical and Social Issues in the Genetic Engineering of Animals* and *Farm Animal Welfare.* He is a principal architect of US Federal laws for laboratory animals.

Dr Mercedita A. Sombilla, *International Rice Research Institute (IRRI), PO Box 933, Manila, Philippines.*
Dr Sombilla took her doctorate in Agricultural Economics at the University of Minnesota, USA. She worked for the International Food Policy Research Institute (IFPRI) after completion of her graduate studies in Minnesota and prior to joining the International Rice Research Institute. At IFPRI, she helped develop IMPACT, the global model for long-term projection of supply and demand of various commodities. At IRRI she currently holds a joint position as an Affiliate Scientist of the Social Science Division engaged in policy related research activities and also as an Acting Head of the Liaison, Coordination and Planning Unit and is responsible for the donor relations activities of IRRI.

Dr George C.W. Spencer, *St Edmund's College, Cambridge CB3 OBN, UK.*
Dr Spencer read chemistry at Imperial College, London and took his DPhil at Oxford University. He subsequently became a postdoctoral scientist at the University of Cambridge where he then read for the Bar. He is currently working in London in the Intellectual Property Department of the Ministry of Defence.

Preface

The science of animal production has suddenly become world-class news. Everyone, everywhere, knows about Dolly the somatic-cell sheep clone and about mad-cow disease (bovine spongiform encephalopathy, BSE). Both events have jumped unexpectedly into the public arena from the science of animal production. Global awareness on these two issues has brought into focus a more general and growing public unease about what animal science is up to these days. People are concerned about food safety, human health and quality of life for humans and animals. Biotechnology applied to domestic and laboratory mammals is increasingly perceived as 'meddling' in life-processes which are threateningly close to humans. The former image of animal science was of a beneficial and benign contributor to society, quietly enabling livestock producers to produce cheaper, safer and better quality animal products. That image has gone. Cloning of mammals, the threatened cloning of humans, the horror and administrative mess of BSE and its associated new variant Creutzfeldt–Jakob Disease (nvCJD) that is fatal for affected humans, genetically modified food and animal feeds and the health risks of inter-species xenotransplants are international news. Parliaments, governmental and inter-governmental commissions are examining some emerging animal science issues and, in some cases, have imposed legal restrictions.

Not only has the traditional image of animal scientists been shattered, but the reliability and ability of science and of governments to safeguard public interests in food and human health is under question. In developed countries, agricultural research is increasingly funded by commerce and industry. Whose interests are being served? What motivates scientists? Can the proven objectivity of the scientific method be extended to the scientist himself? Or do scientists have personal values which affect their work? And what are these values? In the past many animal scientists have been motivated to contribute to increased animal production and productivity within the context of the unwritten code of the 'common good and the decent thing'. Today there is a growing feeling that animal science is more the servant of business and is driven by a value system which evaluates activities only in reductionist and market economy terms.

A new signal is coming from the public; give us cheap and safe food – but not at any cost; and especially not when the undeclared costs include uncertain risks for human health or other components of quality life such as the environment. The public now perceives that new techniques emerging from animal science laboratories, while offering potentially cheaper food, also threaten the biological stability of human life. The new signals reflect a broader spectrum of values and a deeper appreciation of quality of life, of which cheap and plentiful food is only one component. Biotechnology and food safety are but the tip of a much larger iceberg. There are growing public concerns about intensive livestock production, the interface of farming with the environment, animal welfare, genetically modified feeds for animals and the impact of short-term actions today upon the future. It appears that, after all, man is not interested in bread alone but in the whole quality of life today and tomorrow.

Today, animal scientists are no longer simply serving the farmer and, by extension, the supermarket and consumer. We are now servants of, and accountable to, the whole of society. The new questions have moral and ethical components. Will Dolly lead to human clones as well as cheaper food? Will transgenic animals lead to new human diseases as well as organ transplants? Does biochemical treatment of animals make food less safe as well as improving its quality? Are food additives free of health risks as well as improving flavour? Is the abuse of animals ethically acceptable in a civil society? Are genetically modified feeds for animals adequately tested for animal and human safety. Reductionist and objective animal science, which as the handmaid of market economy capitalism, has achieved so much good in increasing the supply of cheap food is today being challenged by broader quality of life values.

For the scientist this new public arena raises new questions too. Do legal restrictions, albeit temporary, mean that science is no longer free to pursue knowledge without supervision? Are there forbidden areas for animal science? Are scientists truly independent of self interest in their work? And can animal science be trusted to avoid dangers to human health and welfare? Should animal science design its own code of practice based upon serving the community, as other professions have done voluntarily? Can animal scientists continue to pursue knowledge wherever it leads them and be exempt from the consequences of applying new knowledge? Do animal scientists need to bother about moral and ethical issues or can these be left to the philosophers, ethicists and theologians?

Animal science also impacts the Third World through development policies, transfer technology and the introduction of more intensive livestock production methods. The model of science boosting food production in the West is a strong argument for more research funding for developing agriculture to support the increasing populations and higher standards of living. The case for global free trade in farm products is partly supported by expectations that biotechnology in the West will put more food products on the world market. But the concerns about the impact of intensive livestock production on quality of life in the West are matched by concerns in developing countries – although the precise fears are different. Many thoughtful leaders in Asia, Africa and Latin America are concerned about the negative effects of western business combined with agricultural science methods which they see as depleting historic cultures, introducing alien values and bringing economic colonialism. Developing countries experience an ambivalence, comparable with that now growing in the West, about the hidden costs of cheaper and more abundant food when it means the breakup of traditional societies and the import of exotic and foreign values. The track record over the last 40 years of livestock development using imported western methods has a patchy history and is a story of mixed success and failure. Many development practitioners and ethicists are now searching for new systems approaches which include local values and expectations as well as good science. The United National Development Programme (UNDP) and the World Bank (WB) now seek to include Quality of Life Indicators in their investments in developing countries. Animal scientists working in developing countries today face values, morals and ethics in the search for bread – but not bread alone.

This book is designed to bring together animal science, ethics and quality of life issues. The authors come from a variety of professional disciplines from all parts of the world. They are close enough to livestock to open some windows enabling animal scientists to see their professional work in the larger perspective of life as a whole. There is a new drumbeat. We must decide if we shall listen and respond. And responding will need first some new thinking. The aim of this book is to stimulate fresh thinking as a prelude to action.

Earlier versions of many of the chapters in this book were first presented at a Special Symposium on 'Livestock, Ethics and Quality of Life' held at the 8th World Conference on Animal Production, Seoul, Korea from 28 June to 4 July 1998, organized by the World Association of Animal Production.

John Hodges
September 1999

Why Livestock, Ethics and Quality of Life?

John Hodges

Mittersill, Austria

THE PURPOSE OF THIS CHAPTER

This introductory chapter identifies the ethical and quality of life issues associated with livestock, views them in the larger perspective of the current scientific, social, economic, political and business environments and thus provides a context for the following chapters which target specific issues. In this way, the chapter offers background, bearings and connective tissue enabling the reader to enjoy and to navigate the book with better anticipation and fuller understanding of its purpose. Definitions of some main topics are given first.

DEFINITIONS OF SOME MAIN ISSUES

Ethics

The moral aspect of behaviour especially the impact of a person's decisions and actions upon other parties: people, society, the environment and the world in general. Ethics are derived for specific situations from moral principles.

Quality of life

The variety of human experiences which together define a lifestyle and measure the value and meaning of life as a whole. Quality of life is not the

© CAB *International* 2000. *Livestock, Ethics and Quality of Life*
(eds J. Hodges and I.K. Han)

same as 'standard of living' which is usually measured by financial income alone. Typical factors which may be included in quality of life assessments are: average gross national product (GNP), education, healthcare, life expectancy and recreation facilities, food safety, climate, environmental factors, clean water and air, density of population, standard of housing, travel facilities, democratic principles, access to wilderness, etc. The quality of life also includes aspects of community. Higher quality of life is found in communities and nations which offer a sense of belonging, care for others, openness of society, shared cultural values and integrated families in contrast with disintegrating or racially divided societies with latent hostility, a society with respect for others, for animals and for the environment, a society which shares benefits and privileges rather than emphasizing individualism and personal rights to the detriment of others. In general, quality of life contrasts with a quantitative description of a lifestyle.

Biotechnology

The new science of molecular biology, together with its associated techniques and applications which are used to modify, manipulate or otherwise to change the normal life processes of organisms or tissues and thereby to enhance their performance for human use. Animal biotechnology means the application of biotechnology to livestock and laboratory animals with the aim of increasing animal production and productivity.

Animal welfare

The care of animals kept in the service of mankind, so that their well-being is provided for, their natural needs are not restricted and their worth and dignity as individuals are recognized. Animal welfare concepts and practices extend to animals kept for food and clothing, for biotechnology products, for research, for work, for recreation and for sport.

Intensification of agriculture

The process of using external inputs for food production, such as fossil oil, electricity, irrigation, mechanization, fertilizers, herbicides, pesticides, financial investment, large-scale methods and equipment to increase production and to reduce unit costs of food produced. Intensive agriculture contrasts with traditional extensive agriculture which has few or no external inputs and is largely dependent upon local human and animal power.

Intensive animal production

The practice of keeping animals in large-scale units, where mass production methods are used, where animals have little or no contact with natural surroundings and where they are subject to management designed to shorten their time to slaughter and reduce costs of production; a feature of agribusiness with an inherent risk of treating animals as disposable resources.

Development

The policy of seeking to raise the standard of living of a community or nation which involves making changes in the socio-economic structures by introducing new material resources, technology, finance, knowledge, training, economic, financial or physical infrastructures, etc. Development is generally offered through aid and financial investment by materially richer countries towards physically poorer countries and, in recent decades, has often focused heavily upon material prosperity to the neglect of local community and cultural values.

Culture

The whole complex of lifestyle, values, traditions, beliefs and practices which are reflected in the ways a community of people behave, live and identify themselves as different from other communities.

Globalization and free trade in world food

The application of market economy principles such as division of labour and comparative advantage, etc., on a world scale to increase output and/or reduce costs of food production, leading to increased world trade in food without tariffs, duties, quotas and other barriers to free market economic practices.

Market economy capitalism

The prevalent economic system in western society which gives pre-eminence to market forces for allocation of economic resources and which allows price to be determined in a competitive market by demand and supply. Capital, including financial capital, land, property and other assets are privately owned and, together with labour, are mobile in perfect market conditions. The system operates under the rule of law which guarantees and enforces business contracts. The human resources involved include owners who carry the risk of a business, managers, workers and customers. The appropriate financial structures and facilities typically include available capital, stock exchanges and banks.

A NEW CONCEPT

Livestock, ethics and quality of life

What have ethics and quality of life to do with livestock? It seems an unfamiliar association of concepts. The editors and authors of this book consider that the changes now occurring so rapidly in animal science, livestock production and commerce link livestock, ethics and quality of life together in a completely new and important way. Most of us are aware that human society is being reshaped. Societies, both in the developed West and in the developing regions, are, in different ways, under powerful new influences. The changes are far wider and more profound than most of us realize. Traditional patterns are being overturned and quality of life is being redefined.

Scientists may not be aware of the role that animal science, livestock production and animal food products are playing in this restructuring of human lifestyles. However, sensitive and informed sectors of the public feel that the emerging changes flowing from animal science today have increasingly formidable effects and potentially threaten the well-being of humanity. These views come not only from so-called fringe groups; the apprehension can be found among thinking people in different sectors of society, in many countries and different professions. The new challenges to society associated with animal science cannot be regarded as trivial or transitory. Of course, one could launch a counter argument that society is changing in any case under the influence of far greater powers and that animal science is merely trailing along with the rest of society. This is not an adequate explanation. Why? Food production is a basic human activity and lies close to the functioning and heart of traditional, modern and post-modern societies. A revolution in food production and distribution affects all societies at deep levels.

The changes in livestock production which raise ethical and quality of life issues can be grouped in three broad categories.

1. Biological innovations including emerging biotechnology.
2. Consequences of intensive livestock systems.
3. Globalization and free trade in food.

The new alliance of animal science and business

The potential impact of these new and emerging issues upon humanity in both developed and developing societies is very great. Technological methods flow from animal scientists and are applied in the market place by business throughout an integrated food chain which now affects all consumers in the West. In the large-scale economies which now direct western society, consumers of food – everyone in fact – are captive to the market sources and outlets ordained by big business. Reaction to this loss of choice is one reason for the growth of the organic food sector in some western countries. The biotechnological and production methods now being used in animal food production are distasteful to many people. The alliance of big business and biotechnology seems to threaten the safety of food, downgrade the place of animals in society and challenge the integrity of basic life processes.

Further, the pressures for globalization of food and animal feed will inevitably bring developing countries into an orbit which is essentially western in culture and values, with negative impact upon their traditional quality of life. Some professionals concerned with development and some leaders in developing countries see that this move could easily lead to a new economic colonialism of the worst sort – because it is directed at the basic activity and lifestyle of people in developing countries – food production. Concerns are also being raised about the vulnerability of populations of the developing world to the use of biotechnology for food production in a competitive market economy. These issues, potentials and threats lead back to two main sources:

- animal science which develops biotechnology and animal production methods;
- agribusiness, especially the food and pharmaceutical multinationals which, during the 1990s, have positioned themselves for a larger share and more control of food resources at the global level.

The alliance of these two sources – animal science and agribusiness – is clear. The independence of the former and the motives of the latter are, therefore, increasingly questioned. This scenario automatically leads to consideration of ethics and quality of life. At root, ethics is concerned with the way in which the actions and behaviour of one group affects the interests of others. How does the new alliance of animal science with market economy capitalism affect the quality of life of people in the West and in developing regions? The alliance is an emerging factor of modern life which affects everyone because everyone eats. Food production and marketing is different from the situation for a product which has to be advertised, where the consumer can still choose not to buy. *Everyone eats*. In this sense, food production is a public issue and should be subject to public scrutiny and influence far more than most other economic activities. Some consider that food is too important to be left entirely to market forces because everyone has to eat.

World population growth

Because everyone eats, this very fact represents a major problem: the world population is still growing and, at best estimates, is unlikely to stabilize until the middle of the 21st century – with perhaps 10 billion people to be fed daily. How can this be done? There are genuine hopes that biotechnology and free trade can make a positive quantum-scale contribution to increase food production and thus avert famine. This book does not reject the need for or possibility of bringing benefits of increased food production by bio-technology and world trade. In contrast, the authors generally recognize the clear fact that such a food–feed programme for human development is not negotiable. However, the authors are concerned with the *way* in which the emerging alliance of animal science and business will implement the new options. Ethics are concerned with objectives, motives, wholeness, health and community. These and other components comprise a quality life – not food alone.

Need for larger and newer perspectives

The current social, economic and lifestyle changes raise new issues which often are not apparent when our world view is confined to one professional discipline. It can be an enlightening experience to hear the perspectives of people working in other disciplines. Human society is varied and complex. We see the whole more clearly when we invite others with a different world view to come alongside. The combined expertise of specialists from differ-ent disciplines brings new understanding.

This book brings together professionals from a variety of disciplines and from all parts of the world: people who previously have rarely thought together about these emerging human experiences. Any serious attempt to face the emerging challenges as society enters the 21st century requires us to listen, to discuss and to be stimulated by those outside our traditional professional boundaries. Whereas some problems in society are amenable to solution by experts in one field, it is increasingly evident today that the emerging problems of animal science and livestock raise issues involving ethics and quality of life that cannot be solved by experts working in isolation, however competent they may be. The new issues are community problems shared by all sectors of society and, increasingly, by all peoples of the world.

ANIMAL SCIENCE IS ENTERING NEW TERRITORY

Since its origin in the 19th century, animal science has given many new techniques to farmers and livestock producers. These fresh approaches generally have contributed to farmers' and animals' welfare, increased production and productivity, placed better quality food in the hands of the consumer and reinforced existing structures of society. Such changes in practice on the farm have been introduced without causing major concern off the farm. Today the situation is different. Animal science is placing in the public domain remarkable new techniques for manipulating the fundamental biology of food animals. The public finds aspects of these novel methods to be threatening for several reasons.

1. The nature of food is itself being changed and people are uneasy about its benefit and safety for human consumption.
2. The biological techniques which work with large mammals can, and will be, applied to humans.
3. Animals increasingly are viewed as disposable resources in large-scale intensive production systems.
4. Biotechnology techniques and new resources for food production are concentrated mainly in the hands of western multinational companies whose first priorities are market share and shareholder value, while the longer term interests of consumers and societies take second place.

A basic question is being raised. What effects are these new animal science practices going to have on society and mankind? Unless new constraints are designed which will protect and provide evaluation of benefits to society, the option of using a new scientific method and the opportunity to make more profit will be the only guiding factors. The sheer ingenuity of new scientific knowledge often results in a philosophy which says: 'If science can do something novel to animals or to humans, let's use it'. There are counter questions which may well be more valid: 'Why use it, simply because it can be done?' and 'What are the benefits and who gains them?'

Questions of this type should be applied to the use of genetically modified (GM) foods increasingly being used in the USA and promoted on the international market at the end of the 1990s by multinational companies and the US government. In the USA in the closing years of the 20th century, rapidly increasing percentages of crop land used for certain food plant species was sown with GM seed. Genetic modification of staple food crops such as maize (corn) and soybeans is usually directed to the introduction into the plant genome of gene(s) resistant to stronger chemical herbicides and insecticides which enable the farmer to use less chemical spray. The beneficiaries are the multinational companies that sell both the GM seed and the chemical sprays, and the farmers whose production per hectare is higher. It is said that the environment benefits by lower use of chemicals but this issue is still in debate as there are risks that the resistant genes can be transferred into related weed species leading to a rotating scenario of higher doses of stronger chemicals in the future and a threat to farmers growing non-GM crops. There is no direct benefit to the consumer or the public. Market forces eventually may reduce the unit price to the consumer, but this is not a foregone conclusion since the supply of staple foods already exceeds demand in the West. A cause of real concern for society is the lack of choice for the consumer as producers of GM and non-GM staple foods have strongly resisted making them distinguishable by label or price. Many people outside the USA, including some scientists, consider that GM foods have been tested inadequately for long-term risk. This example of GM foods illustrates how proponents advocating the widespread use of emerging biological techniques in society tend to invoke economic arguments whereas there are many quality of life issues, of which economic benefits are only one.

Values and beliefs

The comprehensive and real answers to questions of use inevitably lead to ethics and quality of life. Questions about the use or non-use of remarkable new scientific techniques which are capable of intruding into and of changing the basic nature of life processes in humans and animals are deep, fundamental and extend far beyond the supply of food. Such ethical questions cannot be answered adequately without first asking deeper questions about the nature of man, the purpose of society and mankind's relationship with domestic animals. Such second level questions open the door to values, beliefs and traditional practices.

These topics are new territory for western science, which has made remarkable progress since the Enlightenment by focusing upon the scientific method and excluding tradition, religion and belief. In the West during the 20th century, the scientific, modernist orientation of society has driven discussion of beliefs and values from the public to the private sphere. Yet now, remarkably, it is biological science, and animal biotechnology in particular, which is provoking society to debate these topics in public forums, in parliaments and in government, and not only in private or in church. A central question is being posed: is man different from the higher animals? Science can answer that species differences are controlled by genes and that man shares over 90% of his genes with some mammals and, therefore, biologically, is little different from other animals. It is this very similarity which makes mice ideal models for research on human disease, recognizes the value of pig organs for human transplants and raises alarm bells in the public consciousness about cloning. If sheep can be cloned, why not humans? This type of question divides scientists, as shown by the different views of two eminent scientists quoted in the public media shortly after the breaking news on Dolly, the first somatic mammalian clone. Richard Dawkins considers human cloning a potentially beneficial step on the road of evolving human development, whereas Ian Wilmut, constructor of Dolly, told the UK Parliamentary Committee that he considers the technique for cloning humans will be possible in the near future, but that he thinks it would be undesirable to use it. These are non-scientific opinions and are based upon individual ethics and personal views (of scientists) on what constitutes quality of life for humans.

On the questions of ethics and quality of human life, there is a difference between science and scientists. Science is a methodology for exploring, testing, measuring, understanding and changing the physical universe, whereas scientists are ordinary people whose profession happens to be science. Naturally, a person's profession affects his or her world view, but rarely accounts for all a person's beliefs and behaviour, whether he or she be farmer, bus driver, architect, lawyer, home-maker or scientist. Some scientists are atheists, some theists; some scientists embrace the ethics of Christianity, others of Hinduism or Islam; some scientists are pacifist and others engage in horrific experiments on other humans as happened under Nazism in occupied Europe.

Most scientists have long recognized the distinction between their work and the uses to which it may be put. Science is amoral; the uses to which it may be put require ethical decisions leading to beneficial or detrimental uses for mankind present and future, for animals or for the environment. A scientific technique is no different from a piece of wood which is also amoral; but the actions of using the wood to build a house or to club a person to death are considered by most people to be ethically good or evil and by most societies to be legally right or wrong. Ethics determine what you do with scientific technology. Of course, ethics are also involved in scientists' honesty when they state how well science understands a given technique, announce the level of risk and state the possible consequences when its use is being considered.

Are scientists responsible ethically?

A major ethical question now raised in society is whether animal scientists, in their work, can claim immunity from the consequence of possible uses to which their new discoveries are put. It is feasible, as demonstrated by Dr Ian Wilmut, for a scientist to work hard to produce cloned sheep to enhance the benign and beneficial uses of animals for human welfare, while at the same time believing that the technique should not be applied to humans. Interestingly, the ethical division between scientists on the uses of nuclear energy seems often to have resulted in the nuclear scientists who fear the abuse of nuclear power resigning their research roles. Realists, including the scientists concerned, would argue that this does not stop nuclear research; but some nuclear scientists who take the dissident position prefer to put their weight behind anti-nuclear actions rather than engaging in further research. Similarly, there are animal scientists who have resigned their positions because of their ethical views on the use of animals for some forms of research and on the consequences for mankind.

There is little likelihood of science imposing permanent restraint upon itself. Indeed, in the past, academic freedom and security of tenure for professional scholars have been recognized as essential components of a free society. Scientists working in molecular genetics in the early days of recombinant DNA techniques in the 1970s imposed a temporary and voluntary ban on further experimentation until the risks were assessed adequately. However, that event was before commercial interests had identified the profits to be made from biotechnology. Today it is different. The high costs of investment in molecular and biotechnology research result in business interests pressing for quick use. It is also now normal for western governments to monitor, evaluate and decide if and when new products and techniques may be used. Also, governments in the West today take responsibility for placing restraints upon certain types of biotechnology research or upon the use of public funds, as happened in the UK and the USA on research into human cloning after the announcement of Dolly.

The prospect of animal scientists agreeing together, as professional lawyers, dentists and doctors have done, to adopt a set of professional ethics which proscribes certain types of research is unlikely. Historically, society has rarely placed restraints upon scientific research, apart from the normal ethical standards of society. In recent decades, however, animal science research has lost much of its freedom to explore at will simply for the sake of new knowledge, not because of legislation but as a result of changed objectives of funding sources. A major part of animal science research nowadays is both funded and directed towards specific objectives identified by the funding agency, whether it is government or business.

It is precisely this aspect of animal science research which increasingly worries the public. People fear that science in general and animal biotechnology in particular are becoming the servants of big business. Small businesses usually cannot afford to fund much R&D. Big business and wealthy countries are undoubtedly buying many of the best scientific brains. Whereas, in general, the public formerly saw animal scientists as benefactors of society, they now suspect that animal scientists are tied to the profits of business. Consequently, while the image of science as an objective methodology remains intact in the public mind, there is growing suspicion about the ethics of animal scientists.

Ethics and legislation in society

Does western society any longer have an agreed ethic? Western society is increasingly pluralistic. Society is changing rapidly in almost every way: technologically, economically, politically, socially and also ethically. It is therefore a very complex situation in which to consider the ethics of issues which are of public interest. New scientific techniques which appear to challenge the nature of man, animals and human society are not individual private matters. They have to be considered in the public arena. Thus we find a strange paradox. At the same time when parliaments are legislating rights for private behaviour of individuals, for example to choose their own ethics of sexual behaviour, new issues closely related to human reproduction which need ethical evaluation are being placed in the public domain by animal biotechnology.

Parliaments and the law courts are being asked to define ethical standards of behaviour in relation to human cloning, *in vitro* fertilization, embryo transfer, use of frozen semen after the death of the donor (?owner), use of embryo tissue, xenotransplants, transgenic people, etc., and there are many new allied techniques to come. No doubt it would have been easier in the 19th century for parliaments to define and legislate acceptable ethical standards of behaviour for these new techniques as, in those days, society was more stable and unified in identifying right and wrong. Today it is difficult. The old standards of right and wrong, as measured by the law, are no longer sovereign, and laws are being changed. We are forced to go deeper and ask whether traditional values of right and wrong were evolved arbitrarily by society, or whether they had some deeper origin in the morality of good and evil. An even deeper question emerges. Is there an absolute moral structure in the universe? These issues inevitably are addressed by several authors as they examine ethics and quality of life in relation to livestock, biotechnical techniques, culture, development and global trade.

HISTORIC ETHICS IN THE WEST

Europe and North America have been the cradle of modern science and of market economy capitalism. It is therefore helpful to consider the historic origins and practice of morality and ethics in these societies. Values in Europe and North America have been based upon Judeo-Christian principles and teaching. Even after church and state were separated and the political power of the church in Europe and America was curtailed, the moral teaching and ethical behaviour of Christianity continued to be the accepted standards of society. In fact, those standards are the foundations of civilized society in Europe and the USA, which morally and ethically was established as an extension of Europe, and particularly of Britain. The principles upon which the judicial systems are based in these countries are drawn from the Bible even though judicial practice and methods of legislation may differ from country to country.

Western society, embracing these moral principles, has produced legislation over many centuries to define what is acceptable – or rather to define what is not acceptable – behaviour in civil society. Until the second half of the 20th century, it was clear from law and from cultural expectations what society expected of its citizens and of its leaders. People and societies did not always behave ethically but the standards were known.

This is now changing rapidly. Long-established values and behaviour patterns in society are being dissolved and a new pluralism in behaviour and ethics now characterizes many areas of personal life. Since ethics is concerned with human behaviour and particularly with its morality, we speak of business ethics, sexual ethics, legal ethics, medical ethics, political ethics, etc. Society is now being confronted with the concept of scientific ethics – in particular of ethics for animal biotechnology. This is a new issue. Ethics of biotechnology is highly unlikely to mean 'ethics telling scientists what they may or may not research'. It is more likely to be 'ethics stating which scientific techniques now available in society may be used, by whom, when, where and for what purposes'.

Ethics of biotechnology

The scenario of government dealing with ethical aspects of new issues is already found in Europe where the European Union (EU) has a permanent body of advisers on the ethical issues involved in emerging issues. The EU, national governments and parliaments are already engaged in debating and defining whether, where and when GM crop seeds may be grown by farmers; and whether organs from pigs that have human genes can be used as transplants to humans and or monkeys. Around the time of writing this chapter in 1999, the UK government approved and then, following public concern, withdrew its support for the immediate release of GM seeds.

It is a new, difficult and challenging area for legislation – especially in a democracy. On the other hand, dictators, generally with evil agendas, have no compunction in hiring western scientists that have, for example, the know-how to clone mammals and to apply the technique to humans. This possibility emphasizes the point that traditional legislation has been national and applied within the territory administered by the state. Science, however, is not national in application. New international laws, arising from the GATT/Uruguay Round and the World Trade Organization (WTO) and from international conventions such as the Convention on Biological Diversity, call for legal commitments by nation states on issues such as intellectual property rights (IPR). These new international laws are an attempt to regulate the use of biotechnology and international trade and, therefore, ethics and quality of life issues are involved. It is not an easy route to reconcile diverse national interests, resources and stages of development, as shown by the confrontation between the USA and the EU over trade in GM food which is being submitted to the WTO. Another example of difficulty between sovereign states is the USA decision, contrary to that of most other countries in the years following the Earth Summit in Rio de Janeiro in 1992, not to ratify the Convention on Biological Diversity largely due to USA national interests in ownership of biodiversity, IPR and biotechnology products.

Society, science, law and ethics

We may conclude that research and developments in animal science have run ahead of thinking in ethics and law. Society, science, governments and parliaments now face a whole raft of new issues for which the decision-making processes are not clear. Legislation on the use of emerging techniques in animal biotechnology adapted to human medicine is not really susceptible to the process of decision by voting on party political lines. Politicians of any colour are equal in not knowing the answers. They have to be told the nature of the risks, and who can tell them? Only the scientists. Governments today have in-house scientists, but these men and women are not bench scientists – they inevitably become bureaucrats. One of their main tasks is to ask their fellow working scientists for experimental data, to examine and analyse it and then to advise the relevant minister(s) in private on what to do. The method has some strengths, that of independence, but also some weaknesses; the scientists cannot be expert in all fields and are in any case dealing with the data second hand.

Therefore, some governments now use another technique – to appoint a Commission, Committee or Review Board, consisting of both practising scientists and individuals of experience and stature in public life, whose terms of reference are to study all aspects of the emerging field. Their brief usually requires them to listen to everyone with knowledge of or an interest in the subject. Thus, they interview scientists, lawyers, sociologists, business people, concerned individuals and bodies such as non-governmental organizations (NGOs), churches and religious bodies representing the plurality of moral and ethical positions, etc. The findings and recommendations of such bodies are usually, though not always, accepted by governments and may become the basis for law. Examples from the UK and the EU include Reports on Animal Welfare and the Ethics of Emerging Reproductive Techniques.

There is a problem in keeping up with science, whose very nature is to spring surprises. Thus it is a very difficult task to anticipate what new technique will next emerge and cause disarray in the public arena. The debate is often confusing for the general public.

Conflicts of interest occur over how society should handle new biotechnology techniques as they frequently result from huge investments in R&D. The multinationals concerned usually patent the technique and are desperately keen to start production to gain a new market. Their pressure upon governments is great and, with both national revenue and national pride at stake, governments may be tempted to give easy approval before being sure. Governments may seek to persuade scientists to make definitive statements about safety and absence of risk, which every biological scientist knows is nearly always a matter of probability level rather than an absolute statement of clearance. Politicians do not like probabilities. Scientists do not like ethics. Consumers and users do not like risk. Business does not like waiting. The problem is complex, and it will not go away.

Society watches, waits and votes

Today the public debate on these issues means that society, science and governments are re-encountering the moral terms good and evil and the legal terms right and wrong. While the law may say what is right and wrong and thus define what is allowable and forbidden in society, people and sections of society have deep convictions about what is good and evil.

To return for a moment to nuclear power. Most people in the West agree that since nuclear power is amoral it can be used, under proper political control, for purposes of peace or war. Nevertheless some countries, for example Austria where I live, have decided to be non-nuclear countries. The citizens of Austria voted in a referendum against nuclear energy for any purpose. If one asks individual Austrians for their reasons, the most common response is that any use is wrong. It seems that people consider nuclear power as wrong because, even though it can provide clean energy, it is perceived as capable of harming people, animals and the environment.

We may now ask: 'Shall we see countries, regions or communities which decide by referenda to ban biotechnology?' 'Shall we see communities which are against the use of the modern biotechnologies for food production?' The answer is beginning to appear and it is 'yes'. Organic farming in 1999 was practised on 10% of farmland in Austrian agriculture, whereas in France it was less than 0.55%. Other EU countries lie between these levels. In 1999, the EU and especially the UK experienced deep public and national division over GM foods of both crop and animal types. The UK government extended the trial period for GM crops on approved sites by several years and postponed approval on release of GM seed. There is an EU ban on the use of bovine somatotropin (bST) which enables cows to produce more milk. Further, the call throughout Europe for adequate labelling of GM foods in the shops resulted in 1999 in several major supermarket groups refusing to sell GM foods. Thus, we have a situation in which some biotechnology methods and products considered to be safe by multinational companies and their scientific advisers are rejected by sections of society.

It may be thought that public opposition to animal biotechnological practices and products arises solely from concerns about safety for human health and the environment. However, as explained in Chapter 4, the majority of resistance within Europe is based upon moral rather than safety grounds. It is becoming clear that people in Europe do think about practices for humans, animals and for the environment in terms of good and evil and not only in terms of what legislators consider to be relatively right or wrong before the law.

TRUST OR LACK OF CONFIDENCE IN GOVERNMENT AND SCIENCE

The historical background of any society has a major impact upon moral codes and ethical behaviour. We have already noted the Christian and biblical roots of traditional morality and expectations of behaviour in Europe and North America. There are other components of history which affect the way in which a society makes ethical decisions. An example springs from the current differences in views on biotechnology between Europe and the USA due to their recent histories and despite their descent from a common culture.

It seems to be difficult for the USA to understand European attitudes and values in relation to food products deriving from biotechnology. The USA is now putting on the world market milk products from cows treated with bST, food and animal feed made with GM soybean, GM maize and other GM foods. These products are not freely accepted within the EU. A view commonly expressed from the USA is that Europe is using concerns about the safety of these products to protect domestic producers from US competition (Chapter 8). Clearly it is difficult for the USA to understand that there may be motives other than business, which is the dominant component of American culture. The American argument runs that since the USA government has approved these products for its own consumers, the EU must be hiding behind false defences to protect its own market. In fact, there are other valid reasons (Chapters 9–11).

Europeans have strong historic reasons not to place such confidence in governments and scientists as exists in the USA. European society has living memories of a government elected by the people in a democracy which carried out genocide and experimented with basic life processes of enslaved humans through torture leading to death – the Nazi government of the Third Reich. Within living memory Europe has also suffered under Mussolini, Franco and Stalin. Those are strong reasons to distrust governments and scientists allied with governments.

Further, evidence is now emerging in several European countries, including Sweden which has always publicly taken a positive stance on human rights, of forced sterilization of individuals who did not reach certain physical, mental or emotional standards prescribed by the government. These practices were being carried out even as late as the 1970s.

Europe experienced mad cow disease (bovine spongiform encephalopathy, BSE) during the 1980s and 1990s which led to the death or compulsory slaughter of approaching half a million cattle in the UK. Despite constant assurances by the UK government over 10 years that there was no danger of humans being affected – because scientists were certain that there was no pathogen present – the evidence eventually burst upon society that a new variant form of Creutzfeldt–Jakob disease (nvCJD) was affecting humans and that this had come from eating beef from affected cows. The whole of Europe suffered through the agony of the debacle when this truth about nvCJD eventually broke in 1996, a decade after the original cases of BSE in cattle. By the end of 1998, 37 people had died from nvCJD. Since the lead time for symptoms to appear is unknown, estimates of the final numbers of fatalities remain uncertain at the time of writing.

In early 1999 Europe was again the scene of ethnic cleansing in the Serb Province of Kosovo where the elected government purged some of its own citizens from national territory.

The expectations of European societies for ethical behaviour by their governments have been reduced to very low levels. A morality exists in the hearts of the people which is not reflected in the ethical practices of governments, leaders and their scientific advisers. At heart, it seems that people still have absolute standards of good and evil as a basis for behaviour and relationships with other people, with animals and the environment. A deep moral conscience overrides the prescription of laws by governments in which they define what is legally right and wrong. Europeans are clearly more suspicious of biotechnology than the people of the USA and, based upon experience, want more firm evidence over longer periods of time before they are willing to grant it free access to their society.

RELATIONSHIP OF QUALITY OF LIFE TO ETHICS

The advent of rapid travel and instant communication which started in the 19th century and accelerated in the 20th century has had a revolutionary impact upon life everywhere – with both positive and negative effects upon the quality of life. Until the 19th century, communities were isolated. Few people travelled from home more than a few days journey by horse or foot. In earlier and simpler times, the infrastructures of community were relatively stable. They included the local natural environment, the types of crops and animals used for food and clothing, the basic diet, the types of animals used for work, the community values which governed behaviour and the long-standing religious beliefs which usually supplied authority for the values. Very few of these infrastructures were negotiable; and they defined the basic quality of life for everyone. Naturally, in every society, the rich had relatively more goods and power than the poor. However, overall, the pattern of expectation for rich and for poor alike in a community was predictable. Life changed little from one year to the next.

Conflict has always been a feature of human society both within and between communities. In earlier days, disputes were usually about material issues; for example, land, water, grazing rights, cattle, natural resources, colonies, slaves or other booty. Over the centuries, humans have fought against other communities, races and cultures for possession of such material resources with the aim of improving their own quality of life. World War I, also based upon this pattern, was perhaps the last of such wars.

Later in the 20th century, aggression took a new form, with systematic, large-scale and deliberate attempts to change the whole definition of the national lifestyle. These new forms of aggression were based upon novel and attractive ideologies, Nazism and Communism, which motivated the aggressors first to redefine socio-economic lifestyles within their own society, nation or race and then to attack the way people lived in other nations or, as they intended, throughout the world. These two ideologies aimed to change the nature of human society, forcibly reshape the quality of life for ordinary people and indeed to redefine the meaning of life. The leaders knew that they had no chance of achieving this grand objective of a new quality of life unless they abandoned all morality and developed a behavioural ethic suited to their personal aims.

Thus the leaders of Nazism and Communism spoke about the bright new future their ideology would bring for everyone; but, in practice, their behaviour was driven by self-interest and they totally neglected the true interests of those whose lives they alleged would benefit. These ideologies failed – only because of the strenuous efforts of nations still embracing civil values and ethics. However, these evil empires nearly succeeded. Their early success was due largely to their ruthless disregard of all morality and ethical behaviour in relation to others.

History shows clearly that although Hitler and Stalin led new movements which proclaimed a new era with a 'better quality of life', in fact Nazism proclaimed a better quality of life only for select ethnic groups by prescribing death for others, Jews and gypsies, and slavery for Slavs. Soviet Communism proclaimed a 'better quality of life' for all, and although it was successful in breaking the old social system with its many bad features and low quality of life for ordinary people, Communism took away most of the remaining positive aspects of life which people valued, and finally delivered poverty to all except a small elite in most areas of life.

There are two important historical facts for us today. First, in their conviction and determination to redefine quality of life, the leaders of Nazism and Communism were driven by self-interest and discarded ethical issues which would have moderated their behaviour in the interests of others. Second, the leaders suppressed all opposition and listened to no further discussion. These foolish, but for them essential, actions gave them freedom to do what they liked without restraint against people, property, land, environment, religion and historic cultures. We need to recall that these very actions were taken in the name of a positive ideology which undoubtedly gave benefits to the privileged few. Once leaders have no moral or ethical limits and all dissident voices are silenced, an ideology can do what it likes to fulfil the dream of a new quality of life. So both Hitler and Stalin killed millions – and delivered only devastation in every area of life – all in the name of 'human progress'. The world looks back on Hitler and Stalin as evil men and upon Nazism and Communism as evil systems. They had no morals, no ethical standards and no concern for others. They treated humans as disposable: that is why they are seen as evil leaders.

A new ideology – a new dilemma

At the start of the 21st century, the world faces a new dilemma. The high ideal of providing a better quality of life for everyone in the world is immensely important and commendable, as a majority of people in the world still live at primitive and subsistence levels. During the late 18th, 19th and 20th centuries, the West developed an ideology, market economy capitalism (MEC), which has brought higher material standards of living to western society. MEC works: in western society, MEC has progressively increased the average level of GNP. MEC is very attractive to political leaders who seek a solution to the deplorable poverty in which a major part of the world population lives. MEC, combined with science, has proved itself a powerful means of increasing wealth in general; and, within the subsector of food production, new biotechnologies allied to MEC appear to offer new options for increasing food and animal feed production which are very important components of higher quality of life. It is natural that western leaders should advocate the adoption of MEC for other nations and less developed parts of the world.

However, there is an ethical dilemma. The adoption of MEC in western society has changed the socio-economic culture and has redefined the quality of life largely in material terms. Today, the West is a business-oriented and consumer society where the allocation of resources of all sorts – economic, human and natural resources and of time and space – is based upon the values of MEC. Other components of lifestyle traditionally not in the market place, such as education, recreation, healthcare, family relationships, religious activities, etc., are also evaluated in MEC terms in the West.

To people in many other cultures, quality of life in the West appears a mixed blessing. While they admire and want increased material prosperity, they are cautious and have reservations about yielding their whole way of life to the western definition of quality of life. It is here that ethics – behaviour appropriate to the interests of others in given circumstances – has to find a place. The dilemma for the West and for leaders in developing countries who seek the prosperity of their people is to behave ethically and with sensitivity to the interests of others when advocating MEC as a way, as the only and guaranteed way, to improved quality of life (Chapters 9–11).

A NEW WORLD ORDER – GLOBALIZATION

Western leaders, in particular the USA which is the flagship of MEC, propose to bring a 'better quality of life' to all throughout the world. During the 1990s, since the fall of Soviet Communism, several visionary Western leaders have spoken of a 'new world order'. The ideology undergirding this new era is global market economy capitalism (GMEC) and, in the minds of purists, it is usually linked with western-style democracy. GMEC is a global extension of MEC. Globalization, as it is called at the end of the 20th century, is expected to multiply for the rest of the world the material prosperity of the West. The GATT/Uruguay Round and later the WTO are international attempts to provide economic and legal bases to which nation states are invited to subscribe. If you wonder what this has to do with animal science, it should become clear in a moment. Science is the technological powerhouse which drives business today. Biotechnology is increasingly driving the food and pharmaceutical multinational companies. These companies are at the centre of the campaign to bring free trade and globalization of food to the whole world community.

The development of market economy capitalism

We need to pause and examine the track record of MEC in Europe and the USA, which has become the West Point of capitalism. There is no doubt that as MEC grew slowly in the West it delivered better material prosperity for the majority of people. However, MEC has no in-built ethic. In the absence of legal regulation, MEC showed itself capable of limiting or reducing the quality of life of vulnerable segments of society. For example, slavery, slave trading, children and women labouring in the mines, excessive daily work hours, wages taken as rent by an employer/landlord were the offspring of MEC in the early days of industrialization. These practices were outlawed in the 19th century through legislation in Europe and North America by parliamentarians who had moral standards and fought for laws to enforce ethical behaviour in public life. The reformers insisted on civilized ethical standards being adopted as normal social behaviour – enshrined, enforceable and guaranteed by law – which is itself a positive and essential civil structure.

Later, other forms of MEC exploitation were restrained by law, including monopoly powers, work conditions in factories, mines and quarries, health and safety requirements for employees and consumers using the products, hygiene standards for food products, limitations upon loads carried by buses, trucks and ships, and so on. MEC in its raw form has been trimmed in the West into an acceptable mould, compatible with equality and human rights. Without such trimming and restraint, MEC would have taken on some of the characteristics of Nazism and Communism by raising the quality of life only for the privileged, educated, informed, wealthy, healthy and intelligent, and by depressing the quality of life for the majority. Although this thought may seem abhorrent and impossible to the present generation, we have only to look back to the 19th century in Europe and the USA as industrialization and MEC spread, and we can quickly verify the fact that, without legal restraint built upon moral principles in an open society, leaders in any society become ruthlessly selfish. To confirm the point, we need only reflect that a central principle of MEC is self-interest.

Today the West has a form of MEC which is constrained by laws and moderated by social welfare. These legislative restraints are more extensive in western Europe than in the USA which sometimes sees European MEC as soft, and therefore less efficient. Society has to pay for the quality of life it chooses.

ANIMAL WELFARE AND QUALITY OF LIFE

Today, animals can be produced in ever increasing numbers, be endowed with designer genes, pass a fore-shortened life in restricted confinement, be fed controlled rations and finally be processed into ever cheaper meat products for sale on the world market. These animal products give an improved quality of life to many people previously consuming little animal protein. At the same time, intensive livestock systems which increase the volume of animal products on the market and reduce unit costs also raise new concerns about the treatment and welfare of animals. These new concerns are especially vocal in the West where human rights have given birth to the concept of animal rights. In most western countries, governments differentiate between animal welfare and animal rights by passing legislation addressing the former. This legislation is based upon the moral justification to avoid pain and to provide animals with conditions in which they may relax and perform normal functions without difficulty or abuse. At the same time, the legislation seeks to minimize any increases in production costs. The area is a minefield for those seeking precision of definition.

Societies and individuals have various models for the way in which they view animals which lead to a wide variety of human behaviour ranging from vegetarianism to seeing animals solely as disposable biological resources. These models and associated issues are discussed in detail in several chapters.

The ethical dilemma concerning quality of life is well demonstrated by the issue of animal welfare. On the one hand, in the West, it is often pointed out that a society which abuses its animals is less civilized and therefore offers a lower quality of life. On the other hand, the availability of animal products at lower costs enables lower income groups in western society to buy them and offers the prospect of animal products to the billions of people in the developing world to increase their intake of animal protein and thus experience a higher quality of life.

ANIMAL SCIENCE AND FOOD PRODUCTION

The new world order will greatly affect global food production and world food trade. Animal biotechnology is a major player in this new scenario. For 100 years from the mid-19th century to 1945, many changes were brought about in livestock production as science was applied mainly by farmers and livestock owners, freely tutored by government extension agents into positive husbandry practices accompanied by benign animal welfare. During that period, animal science was directed mainly to the farmer and not to the upstream farm supplier and downstream food-processing businesses. Science increased the production of cheaper food, usually enabled the farmer to make a slightly improved profit, and improved the quality of life for the whole rural community and for the urban consumer.

During the 50 years since the end of World War II, farming in the West has changed from a way of life to a business. Animal producers now make decisions as business people husbanding their profit, return on investment and costs. As the benefits of scale are captured, livestock production has intensified. Science serves the farmer less directly. Today, research impacts the farmer mainly through upstream and downstream businesses, especially the chemical, food-processing and pharmaceutical companies. Many of these businesses merged in the 1990s, so that at the end of the 20th century a few multinational companies are enormously powerful players on the world scene.

Conflicting effects of globalization of food production

The ideology of GMEC says that business, big and small, will bring cheaper and larger quantities of food to feed the still rapidly growing world population. Science, as the handmaid of business, particularly big global business, is contributing powerfully. There is no doubt that good science allied to large-scale business can cause the earth to be more fruitful. The threat of Malthus can be averted again. It is difficult to visualize how the expanding world population will be fed without a new technological breakthrough. Biotechnology appears to offer the prospect of replacing the Green Revolution which is now running out of steam. However, biotechnology is not yet tested and proven. The risks have not yet been fully assessed. The precautionary principle being debated in the inter-governmental forum of the International Convention on Biological Diversity is designed to provide protection against premature use. There are nations for and against the definition and international administration of the precautionary principle and, regrettably, the strongest voices against come from the West.

Countries with high population density, little good farm land or limited water – or countries with burgeoning populations in Asia, Africa and Latin America – could give up their dependence upon domestic production of some foods and buy imported foods on the world market. The difficulty faced by many rural segments of society in developing countries is lack of cash to buy imported food. The proponents of biotechnology have an answer. Farmers in developing countries can buy GM crop seeds and, in time, GM livestock, semen or embryos to increase their domestic food production.

The ethics of the multinational companies are emerging as a new factor in world trade. A subsidiary company of the Monsanto multinational has developed a transgene which prevents germination of second generation GM seeds. Thus the farmer cannot save seed from the first year crop and has to buy new seed. The transgene, known as the terminator gene, is deliberately inserted together with the other transgenes to provide greater financial returns for the company which has invested in the research. The patent of the terminator gene is held in the USA by the Monsanto Conglomerate and the US Department of Agriculture. The concept and intention has appalled some other authorities in this field – for example, the government of India has banned the import of seeds with terminator genes.

While the new world order offers potential increases in prosperity, it is clear already that in the food sector the benefits potentially flowing from the new options of increased international food trade and biotechnology are subject to the ethics of those controlling the resources.

The ethical dilemma

Division of labour does work. The law of competitive advantage does work. MEC does work. Economies of scale can be gained. Larger quantities of food can be produced when specialization of labour and science are introduced, as demonstrated in the West over the last two centuries. Malthus's predictions can be averted and society can become wealthier. There is no doubt that as mankind enters the 21st century, the application of the same economic principles combined with new science on a global scale offers the only real prospect of feeding the world. This book does not disagree with these prospects. Rather, the authors are concerned with the ethical and quality of life issues which are intimately entwined with the application and outworking of this socio-economic system to food production.

The proposed new world order involving globalization of food production is basically an amoral system. The quality of life which the new system produces for the various parties involved depends on how the system is structured and whether it is subject to the rule of civilized and humane laws designed to produce equitable results. The existence and observation of such international and national laws depends, in the end, upon the ethics, motives and objectives of the leaders. The organization of economic resources using these socio-economic laws, whether at local, national or global levels, is in the hands of leaders. The most influential and controlling hands are those of leaders and decision-makers with existing access to financial capital, scientific knowledge, technological methods and products, and market intelligence. Parties who contribute only labour and then comprise the majority of the consumers can also benefit, but the sharing of the increased wealth and of the risks depends upon the ethical decisions of the leaders who have pre-existing resources.

The need is for ethical leaders who understand the 19th century history of MEC in western countries which created more wealth but also widened the gap between the richer and the poorer segments of society and contributed to a just and civil society only when national legislation was enacted.

ETHICAL LEADERSHIP OF INTERNATIONAL CALIBRE

Leadership of the new global approach to food production rests with governments of sovereign nations and with the multinational companies. Attempts are being made to introduce new international laws for trade through the WTO and for biotechnology and biodiversity affecting food through the International Convention on Biological Diversity. Unfortunately, these new forums at present are scenes of conflicting agendas. There are divisions between North and South, between sovereign nations with or without biological and intellectual resources, between countries with and without biotechnology resources, between nations having food surpluses and the potential to produce more and those nations in food and production deficit and growing populations.

The divisions which surface in these international fora are *not* about economic principles or science itself, since all agree there are benefits to be gained by increased world trade and that science and biotechnology can produce greater wealth. The differences are about sharing, motives, objectives and control of the market. National self-interest is stronger than any other item on the agenda. Difficult issues are: biosafety, farmers' rights, intellectual property rights, breeders' rights, patenting, access to natural biological diversity, prior informed consent and the precautionary principle in the use of new knowledge. The debates revolve around such questions: Who is to benefit? Who is to control the increased food? Who owns and has access to technological know-how? How are uninformed farmers and consumers in poorly developed areas of the world to be protected against exploitation? Who pays for intellectual property rights? Can genes be patented?

These are ethical questions related to changes in the quality of life at a world level resulting from wholesale GMEC. In summary, the basic questions are:

- What will the application of these principles of GMEC on a global scale do to people, animals, the environment?
- What sort of people shall we be if we give ourselves whole-heartedly to this ideology?
- Does the ideology offer only advantages?
- Will it bring better quality of life to all, or only to some?
- To whom? And who will decide?

Self-centred interest has to yield to a new sense of purpose and public interest in promoting prosperity for the world community as a whole. Arguments about sharing the cake lead nowhere, if there is no cake to share. Unjust and unethical legal arrangements may force cooperation on non-beneficiaries for a short time. However, in the longer run, prosperity and peace result when all benefit. Partners are more productive than slaves or coerced business partners. Contracts and law have to be just and equable to survive. War over food is a real danger in a world with 10 billion people, and will benefit nobody. Finally, can the materially rich 20% of the world claim that they have a high quality life when they know that 80% of their fellow human beings experience a decline in their relative quality of life? Is it ethical to ensure welfare for one's livestock while fellow men and women in the 'globalized village' are unable to enjoy the basic elements which define human life?

Dangers of only one ideology with no opposition

It is appropriate briefly to recall the promises of the two earlier 20th century world-scale ideologies which proposed to change the socio-economic and cultural shape of the world. The leaders of Nazism and Communism were so convincing that they hoodwinked millions of people. Nazism gained political power legally through a democratic election; Communism through civil war – not by conquest from outside. In both cases, ordinary members of society supported the new leaders and expected the ideology to deliver a better quality of life. Communism captured the support of many intelligent people in the West for decades, and many leaders of developing countries enrolled their countries in Communism – for personal rather than national benefit.

Now that it is all over, we must learn the lessons. A major lesson is to be suspicious of leaders in a position of power without an opposing ideology and who advocate a new ideology for others. Rarely are the rich and powerful truly interested in the welfare of the poor and weak, except as a by-product of their own search for more wealth and power. Often their selfish agendas are cloaked by arguments about the benefits which will flow to others if their ideological plans are adopted. If there is one lesson to be learned from Communism and Nazism, it is that leaders without a valid opposition constantly checking, monitoring and challenging them are likely to become selfish despots. 'All power corrupts and absolute power corrupts absolutely'. Based upon this interpretation of history by Lord Acton, there is legitimate reason to be cautious of those who promote GMEC from a position of wealth and power, especially when their first agenda item is national prosperity to be gained through globalization. The leaders of the food and pharmaceutical multinationals are westerners in origin or in education and values. They represent their shareholders. Their aim is increased profit for the company. This is all legitimate within the global market economics. The interesting question arises when the ethics of being a good leader for your company conflict with the best interests of the customer. When a new business ethic is needed, can it be provided by the leaders of market economics or, as with the growth of MEC in the West, are legislative controls the only way to protect the weak? If international laws are needed, who is to provide and administer them? These are new tough questions for all parties.

Summary of the ethical issues in the global market place

Since the issues outlined here may be easily misunderstood, let us restate them briefly. MEC supported by science has proved to be an enormous socio-economic success in the West for producing increased wealth. This is not in dispute. The key issue is to ensure that GMEC is equitable. In its unrestricted national form, MEC has always needed regulation beyond the automatic 'hidden hand' of restraint due to self-interest which Adam Smith expounded. Extending MEC to a global level beyond the reach of national legislation will also need regulation.

Again the world has seen in Russia and the New Independent States (NIS) in the 1990s what happens when MEC is introduced into a lawless society. Very, very quickly a small percentage of the population, usually those with connections and pre-existing resources gained from relinquished Soviet government assets, can gain control of major sectors of the economy. In Russia at the end of the 20th century, being a 'Russian Biznizman' means to the average citizen that you are ruthless, powerful, wealthy beyond measure and totally disinterested in ethical behaviour. The principles of MEC, including privatization of state industries and assets, has been strongly advocated by the West as the only way forward. Doubtless it is the only option – but, in the absence of the rule of law and enforceable contracts, MEC has again proved itself to be a means of exploitation.

The questions remain. Who is to legislate and to enforce restraint on biotechnology and on global market economy practices to avoid exploitation without at the same time taking away the power of the system to perform effectively? These are ethical questions which affect the quality of life of the billions of food consumers in the world, especially those whose present quality of life is minimal or below minimum by western standards. These issues are considered in detail by some of the authors in this book. In the meanwhile, this chapter closes by identifying that ethical leadership is needed both in the West and in the developing world.

The need for ethical leadership

No intelligent and thoughtful person would deny that there will be risks, to people, to animals and to the environment, if the powerful combination of MEC plus animal biotechnology is allowed to develop without restriction. Equally, all objective assessment predicts that the quality of life will change in all the communities involved. The major problem arises when the leaders and decision makers – those with knowledge, education, capital, access to political power and the control of existing businesses – make decisions which will redefine the quality of life for others without power whose values, traditions and cultural assets will be changed.

The proponents of GMEC argue that their ideology will bring increased material prosperity to all. Doubters think that the increased wealth will widen the gap in the world between the rich and the poor. Others simply say that, while they like the increased material prosperity, they do not want to be westernized. For them, it is a question of the values defining quality of life.

At the end of the 20th century in the West, with our individualism supported by democracy and law and powered by a long period of successful science plus MEC, society has largely defined quality of life in material terms. However, a new realization is now appearing and a searching for other components of the good life, especially for those components which in the past were part of the Commons. Purist economists might argue that everything, even the Commons can be bought and sold – and to some extent this is true. The financially well-off buy their homes and recreations in unspoiled locations and eat better quality, gourmet food. The redefinition of quality of life has triggered a new search for ethics in western society – a search stimulated by the emerging biotechnologies.

It may be that the search will lead towards new law – even to the acceptance of new, equitable and enforceable international law. However, good law needs informed, ethical and inspired leaders for its creation and for its just application in society. Increasingly in the West, we see the law and the legal system being distorted and manipulated to exonerate those who abuse the intent of the law. Wriggling through loopholes to escape the intent of the law is unethical, but legal. Such legal wriggling may enable leaders to avoid conviction for doing wrong. The world today is looking desperately for leaders of moral stature in all areas of society, in the West and in the developing regions.

Today, western society is examining and questioning the ethics of its leaders and decision-makers as never before in all areas: political, business, financial, economic and in science. Society in the West and in the developing regions is looking for quality leaders who value the difference between good and evil better than they value expediency. In addition, society wants leaders for whom doing good and not evil is more important than being able to escape from the wrong side of the law. In the end, that is what ethics and quality of life are all about.

Animal Biotechnology: Convergence of Science, Law and Ethics

2

R. Brian Heap[1,2] and George C.W. Spencer[1]

[1]St Edmund's College, Cambridge, UK; and [2]The Babraham Institute, Cambridge, UK

INTRODUCTION

The history of agricultural progress in the UK presents a telling case of scientific and technological advances, innovation and government policies that resulted in an explosive mix which profoundly influenced food production after World War II. Productivity showed a surging rise in output whether measured per hectare or per animal, with recorded growth rates of about 2 or 3% per annum. Looking to the future and the demands imposed by population growth, questions have been posed about whether comparable advances could be achieved since this would be relevant for food security not only in the UK but also elsewhere.

The agricultural industry has become well known for its ability to respond quickly to new scientific breakthroughs, due in large measure to the vision and skills of individual practitioners many of whom are among the best informed of any industry. The UK Government's Technology Foresight Programme for Agriculture, Natural Resources and the Environment identified those areas where promising new targets exist (Hillman, 1995). High on the list were novel biotechnological products for industry (pharmaceutical, animal and plant health, energy and polymer industries); new breeding strategies to improve food security, safety and quality; intelligent systems for precision agriculture; and better horizontal coordination in food production between the knowledge base, retailers and health-conscious consumers, and improved vertical coordination between producers, processors and retail outlets. The speedy incorporation of new technologies was seen as a way to remain at the forefront of production and processing. Yet the adoption of many of these initiatives has been limited by gaps in fundamental scientific knowledge, levels of investment to turn knowledge into technology, overregulation that diminishes competitiveness, and a reluctance to address global, rather than regional food needs and opportunities. Of major concern, however, has been the emergence of public disquiet about new technologies, and in particular biotechnology as it relates to livestock and food production.

BIOETHICS, A GROWTH INDUSTRY

Epoch-making scientific advances particularly in the life sciences now attract a far greater ethical analysis than was previously the case. The brilliant work of Watson and Crick in the early 1950s which unravelled the nature of the genetic code was heralded with universal acclaim. However, in the 1970s, the development of gene splicing techniques for transporting foreign genes into bacteria provoked a 2-year moratorium self-imposed by the scientists concerned. In 1980, *The Recombinant DNA Research and Development Notification Act* was passed in the USA, and the idea of local, state, national and international participation in the regulation of genetic engineering was born. More recently, the National Bioethics Advisory Commission (NBAC, 1997) of the USA was instructed to undertake a thorough review of the legal and ethical issues associated with cloning and, as a result, a Bill has been submitted, but not yet enacted, to prohibit the extension of cloning to human-kind.

Such developments have led philosophers such as David Kline of Iowa State University to ask what it is about biotechnology that makes it special from the moral point of view (Kline, 1990). Biotechnology, which embraces the use of living systems for the purpose of wealth creation and improvements in the quality of life, has proved both intellectually exciting and economically promising. Other revolutionary technologies including microelectronics and superconductivity have been all but ignored compared with biotechnology in the ethical, social and economic debate. The Technology Foresight Programme referred to above has created a means for fund-holders to target their respective resources, and Kline has argued that, as a result, we all have a stake in the outcome. Such changes 'jar scientists from naive political isolationism' into active participation or antagonism. Biotechnology, therefore, has attracted special attention because it directs our attention to key questions about who owns science, who formulates its ethical framework and how is the common good best served in respect of the environment, animal welfare, food safety and food security?

FOOD SECURITY

Reverend Thomas Malthus (1798), an ordained minister of the Anglican church and a Fellow of Jesus College Cambridge, stated in his famous essay nearly 200 years ago that 'population, when unchecked, increases in a geometrical ratio. Subsistence increases only in an arithmetical ratio'. Since 1950, the human population has doubled, and UN projections indicate that it is set to reach about 8 billion by the year 2020 and 9.5 billion in 2050. The trajectory of the sigmoid model predicts that the current exponential increase will stabilize around this figure, though in some models the figure is even higher. Whichever model will apply in the future, global population growth will be checked somehow either by fertility control, socio-economic factors or because of unsustainable consumption levels.

To date, the fear of population outrunning food output on a global scale has not been realized (Sen, 1995), and this has raised questions about whether there are real threats to food security. Dyson's (1996) analysis shows that during the 1980s world cereal output per capita plateaued due to a decline in production in North America/Oceania, a major cereal area where much more than 1 ton of grain per person is produced each year. In southern Asia, the Far East and Europe/former Soviet Union (FSU), where over 70% of humanity live, production per capita was appreciably higher in around 1990 compared with 10 years previously, and output has continued to rise in Europe/FSU. On this analysis, the Malthusian pessimism that population has outpaced cereal production in all the main world regions cannot be sustained. In Africa, however, food production has failed to keep ahead of population growth. The growth rate of the population has been far in excess of the rate of increase in indigenous food production or its potential for improvement within current agricultural strategies. A decline in food output in sub-Saharan Africa has also led to a decline in entitlements since reduced income from growing food has not been counterbalanced by increases in non-food industrial outputs (Sen, 1995).

The Brundtland Commission (1987) defined food security as secure ownership of, or access to, resources, assets and income-earning activities to offset risks, ease shocks and meet contingencies. In other words, not everyone is intended to be a subsistence farmer, but everyone must possess the means to acquire an adequate diet. For most of the world's population this is a rational interpretation of food security, with the prosperous producing that which is surplus to indigenous needs and the less developed areas benefiting from its distribution to areas of scarcity. If increased food output is envisaged, it is important to note that much of the land that is suitable for cultivation worldwide is already in use (11% arable, 24% grassland, 31% forest and 33% polar, desert, stony and rocky land in mountainous regions). If a further 4% was to be utilized by the year 2000, a comparable area could be lost by soil degradation and desertification arising from improper management and climatic conditions. Spare capacity in terms of land usage would seem unlikely to meet the demands of increased production.

Livestock

People replace plant foods with animal foods as material standards of living rise and, as a result, there is an increase in feed grain production, particularly when pasture and rangelands are limited. Overall, livestock products have been estimated to account for about 30% of the total value of food and agriculture. In Europe, the average citizen consumes about 94 kg of meat annually (43% pork, 23% beef and 20% chicken), 13 kg of eggs and 98 kg of milk. Animal products enrich the human diet with protein, bio-available micronutrients (Fe, Cu and Zn) and vitamins (A, D and B_{12}) which are particularly significant for the fetus and growing young, especially in less developed countries where anaemia remains the most serious endemic disease. Current recommendations for a healthy diet in more developed countries are directed at a reduction of red meat consumption, and will lead to decreases, rather than increases in present levels of consumption. This will be more than counterbalanced by the substantial increase contributed by those countries in transition. This is exemplified by Asian-Pacific countries where the share of the world's meat and milk production has increased to about 32 and 23%, respectively. The region also accounts for 46% of the world's fisheries production and has a monopoly in aquaculture (83%).

Whereas vegetation represents a substantial component of the planet's biomass and provides the major source of human nutrition, the direct and indirect contributions of livestock should not be underestimated. Livestock have a special role in many communities. They make effective use of large areas of land that are not capable of producing crops for direct human consumption, such as sparsely scattered vegetation and crop residues. They provide a large component of the fertilizer without which soil productivity would be quickly diminished in much of the developing world's agriculture. Certain species supply draught power for the cultivation of crops, transportation and valuable non-food products such as wool, hides, bones and dung for fuel in poorer communities. Together with their offspring, livestock provide a 'natural bank' of cultural and cash insurance against penury resulting from natural disasters such as drought. In some areas, they are virtually the only viable means of providing a livelihood for the existing population. Important advantages also derive from the integration of livestock and crops where mixed farms are more sustainable than monoculture, though the latter are less favoured in more developed countries where the emphasis is on intensification. Producing animal protein at a global level, however, demands high inputs of photosynthetic energy and it increases pressure on marginal grain-growing land and accelerates soil degradation, neither of which are sustainable at modern levels of human diets based on meat and meat products (McMichael, 1993).

Estimates have shown that animal feed accounts for about one-quarter of all cereal grains consumed in the more developed countries in the year 2000, compared with about one-sixth in 1980. About one-half the grain produced in Europe, North America and Russia is already used as feed. In addition, large quantities of soybeans and coarse grains are sold as feed grains to the richer countries when export-orientated agriculture begins to increase in countries afflicted by serious debt. Unsurprisingly, Rifkin (1992) has argued for the abolition of systems that produce vast numbers of livestock in response to market forces and in disregard for their environmental impact. He states that 'there are currently 1.28 billion cattle populating the earth. They take up nearly 24 per cent of the land mass of the planet and consume enough grain to feed hundreds of millions of people. Their combined weight exceeds that of the human population on earth'.

Taking into account the estimates of costs of feedstuffs and fossil fuels for livestock compared with crop production, Rifkin's arguments may at first appear valid. Animal production is less efficient than that of plants in the conversion of solar energy into components of the human food chain (Spedding, 1984). These comparative advantages are greatly reduced, however, when account is taken of the time-honoured role of livestock in harvesting solar energy and nutrients from the major non-arable areas of the world's land surface including hill, marginal, range or wetter areas. These areas are more suited to grass than to cereal production.

EMERGING OPTIONS

Just as the world could not feed itself today with the farming methods of the 1940s, so farmers can hardly expect to meet the increased global demand in 30–40 years time with their present methods of producing food. Without another agricultural revolution, the fate of the peoples of the less developed economies especially looks grim. New approaches are needed if food security is to be addressed realistically for a world population of 8 billion or more. Against this background, scientists have argued for fresh thinking about how animals and their products could contribute to future food security. Biotechnology has become a potent source of new approaches to food security, whether in terms of animal and crop production or the processes by which value is added to the different outputs. What are the prospects that it will help – would it be feasible to handle the economic and social consequences of a large-scale switch to these advanced technologies even if they make certain farming practices redundant, or will the options offered by biotechnology prove unacceptable because of their impact on animals and the human food chain?

Seven major areas will be reviewed briefly and critically to illustrate how biotechnology has started to provide new options for the systems of husbandry and production. They will also highlight the nature of problems that have emerged for the animals themselves and the extent to which issues of public concern have become prominent.

Genetic selection

Studies of livestock genomes using molecular genetics have increased our understanding of the nature of genetic variation at the level of individual genes, and progress has been made in the development of low-resolution genetic maps. The maps are based on microsatellite loci covering the majority of the genome for pigs, sheep, cattle and poultry (e.g. the European PiGMaP, BovMaP and ChickMaP). The aim is to consolidate into summary databases information from genome mapping programmes in North America, Europe and Australasia and to identify conserved syntenic groups of genes (Andersson et al., 1996). The work is expected to provide more robust forms of marker-assisted selection of valuable traits. It will be limited, however, by the fact that many commercially important traits are probably governed by linked genes located at several sites in the genome.

Diagnostic procedures are also being advanced to aid genetic selection. For example, the screening of genes that code for variants of a fatty acid-binding protein has allowed for more selective improvement of meat quality in pigs. Negative selection has been used for undesirable traits such as the gene associated with stress-related deaths in pigs and with malignant hypothermia. The relevant gene was found to be located within a quantitative trait locus (QTL) for performance-related traits including carcass lean content. For this reason, the gene had been positively selected because of the production preference for lean carcasses. Selection for disease resistance is another long-term aim, and the ChickMaP Project was set up to locate genes that influence resistance to infectious diseases (Burt et al., 1995). A caveat, however, is that on the basis of the past history in livestock and other species, the application of QTL to animal breeding schemes needs to be treated with caution since discriminating between the effects of multiple linked genes and those due to a genuine single QTL will prove difficult. Some QTL effects may not be replicable and associations between alleles at several linked loci may be short lived. None the less, improved knowledge from the genetic maps and the use of genetic markers has been seen as one way to understand and influence naturally occurring variation (Haley, 1995; Georges, 1997).

Biodiversity

An important application of molecular genetics is in the preservation of biodiversity. Estimates show that of the 30 million or so species of living organisms on earth, less than 15,000 are birds and mammals, and of these about 30 species are husbanded for the production of food and agriculture. With increasing population pressures on the environment, diversity is essential if provision is to be made for future needs influenced by events such as climate change. During domestication, separate and genetically unique types or breeds have been developed to suit the local climate and community, which has resulted in about 4000 breeds (Hodges, 1990). These form our primary animal genetic resource for food and agriculture. Based on production criteria alone, the Holstein cow produces the most milk per day, the Merino sheep the finest wool, the Large White pig is the most numerous of pig breeds, but there is great debate about which are the most sustainable and efficient breeds and which will harmonize best with present and future environments. Erosion of domestic animal diversity, however, is a continuing threat. China contains more than 250 breeds of animals, of which about 60 are pigs, but if it was decided to limit activities to about four or five breeds, loss of genetic diversity could be even more serious. The management costs required to maintain the existing genetic pool are negligible compared with the massive cost involved in artificially making a breed to satisfy a specific change in the environment in a form that is sustainable and stable. The Food and Agriculture Organization of the United Nations has established a molecular-based programme to expand knowledge of genetic diversity in each domestic animal species and to enhance existing initiatives in India, Brazil, USA, EC, Latin America and Scandinavia (Hammond, 1993).

Health

Advances are notable in the diagnostic use of monoclonal antibodies to monitor fertility and infertility and to characterize a wide range of bacterial and virus diseases including foot-and-mouth, rotavirus, bovine herpes virus, rinderpest, trypanosomiasis and *Brucellosis*. New molecular vaccines are under investigation for conditions that include the control of temperate and tropical zone diseases (e.g. *Babesia*, *Boophilus microplus*, rinderpest, try-panosomiasis and helminth infestations) and the detection of contamination in cattle and poultry meat from common strains of *Salmonella*. Success has been achieved with the production of a feline leukaemia vaccine, the first product of genetic engineering to combat a retrovirus. Vaccines may eventually include DNA-based vaccines, now in the early stages of human clinical trials (Donnelly *et al.*, 1997), to help circumvent some of the serious problems that arise from chemotherapy resistance and, thereby, reduce dependence on chemicals. In the meantime, livestock diseases remain a source of serious deprivation and hardship for animals and those who depend on them, with costs worldwide running to more than $1000 billion. In the USA alone, food animal health products are set to double from $2.42 billion in 1994 to $4.46 billion by 2001, growing at a 9% compound annual rate. Pharmaceutical products are expected to rise from 36 to 50% over the same period in overall market revenues, while perceived trends in feed additives have been predicted to decline from 53 to 42% and biological agents from 11 to 8%. By now, however, few underestimate the complexity and difficulty in bringing competitive new treatments such as molecular vaccines to the market, for they need to be safe, highly efficacious and less costly than existing treatments.

Production and efficiency

Improvements in the efficiency of animal production continue to be a long-term aim of many research programmes, though the strategy is not without its critics. Techniques that enhance the utilization (and consequently conservation) of feed illustrate the point. They include the use of recombinant growth hormone and related growth factors. Growth hormone (GH) has earned the reputation of being one of the first products of the biotechnological revolution, an accolade marred by controversy.

Frequent injections of the recombinant hormone provide the nearest animal equivalent to enhanced 'nitrogen fixation' since it improves the volume of milk secretion and output of protein by an efficiency gain. In dairy cows, the results from a 2-year post-approval study of bovine somatotropin (bST) showed that it was safe with only 1438 reports indicating adverse reactions in treated cows, or one to eight reports per million doses sold. By the end of 1995, almost 20 million doses had been sold in the USA, and there was no increase in the amount of antibiotics used for treatment of mastitis, an alleged welfare problem according to some earlier reports. Introduction of the GH gene into the germline by oocyte microinjection has proved effective in the stimulation of adult growth rate. An alternative approach has been to use a DNA construct coding for growth hormone-releasing hormone (GHRH) in a suitable vector. When injected into skeletal muscle of mice, it increased GH levels three- to fourfold for up to 2 weeks (Draghia-Akli et al., 1997). The prospect of using this gene enhancement approach in food animals, however, seems impracticable. Better opportunities exist with transgenic fish in which GH gene constructs introduced into the germline enhanced substantially the quantity of protein anabolism (Jun Du et al., 1992; Devlin, 1997). In transgenic sheep, a growth factor construct coding for insulin-like growth factor 1 (IGF-1) has improved wool production. The construct consisted of a mouse ultra-high-sulphur keratin promoter linked to an ovine IGF-1 complementary DNA (cDNA). Clean fleece weight was increased by 6.2% in transgenic animals compared with non-transgenic half-sibs (Damak *et al.*, 1996a, b). Many of these examples illustrate the proof of concept, but it is only in the case of recombinant bST that an expanding commercial product has so far been realized, albeit in a limited number of countries. Notwithstanding its potential importance for incorporation into existing systems of husbandry in less developed countries, socio-economic and socio-political factors have led to a tardiness in the uptake of application, for reasons that will be addressed later.

Feed

Improvements in the utilization of animal feed have been sought through a range of experimental studies. The benefits derived uniquely by ruminants from the bacterial degradation of plant structural carbohydrates to volatile fatty acids result none the less in energy lost in fermentation which reduces the efficiency of this process nutritionally for the host. Better digestibility of low-quality feed has been achieved by the addition of genetically engineered microorganisms or enzymes (e.g. β-glucanase treatment of barley fed to pigs and chickens). In addition, rumen microbial populations have been modified for the same purpose by the introduction of improved genetically engineered bacteria. After re-colonization of the rumen, they enhanced the overall capacity of existing populations of carbohydrate-degrading organisms. Other more esoteric approaches have included the computer-aided design of a protein enriched in essential amino acids. The designer protein was produced subsequently by the introduction of an appropriate gene construct into a suitable bacterial expression system. The formation of hexoses from cellulose and hemicellulose in the small intestine of mono-gastric animals has been attempted by the targeted expression of a bacterial cellulase gene within the exocrine pancreas (under the control of an elastase enhancer), leading to the secretion of the digestive enzyme into the small intestine (Ali *et al.*, 1995). Other studies have aimed to modify livestock metabolism by hormonal administration (steroids, β-agonists) and to improve body composition by stimulation of the adaptive immune system (antibodies to fat cell membranes), but so far they have not pro-gressed beyond the research and early development phases (Armstrong and Gilbert, 1991). Hence, the practical application of these options has so far been confined to improvements in feed utilization prior to ingestion.

Reproduction

Specialized reproductive technologies continue to provide new opportun-ities to improve the consistency and safety of products from genetically valuable food animals and to safeguard rare species. This has been achieved by the dissemination of elite stock, and it is anticipated that the industry will be dominated increasingly by progeny-tested bulls screened for genetic markers of economic traits such as lean tissue growth, low fat products and, in the longer term, specific forms of disease resistance and improved health status. The rapid dissemination of high-quality genetic material from breeding stock has depended largely on artificial breeding techniques, and these are being expanded to the use of sex-selected semen.

Currently, there are attempts to make full use of the reproductive potential not only of males but also of females by nuclear substitution (transfer) procedures. Nuclear substitution is based on pioneering work in Amphibia, but so far the technique has given low rates of development for reconstituted eggs. Embryonic stem (ES) cells, and now somatic cells, promise an ideal source of karyoplasts, and these could provide both large numbers of genetically identical offspring and a way to introduce new genes into the germline (Wilmut *et al.*, 1997). If the technical barriers can be overcome to improve the efficiency of these procedures, such technologies could be turned to advantage in the conservation of genetic material by the *in vitro* production and cryopreservation of embryos, providing a method rapidly to restore populations decimated by disease outbreaks (Heap and Moor, 1995). Whereas the new options from research in reproductive physiology are scientifically promising, the prospect that they will have an early impact on food production other than in niche areas is regarded by many as overly optimistic.

Niche markets and added value

Product diversification and adding value to agricultural outputs by the use of gene transfer techniques in animals have been well demonstrated in the past decade. Selective transfer of genes both within and between different species has been used skilfully to modify the milk of dairy animals, and the potential market value of some candidate proteins in the USA alone is substantial (Mercier and Vilotte, 1997). Over 50 different proteins have been studied, in some instances to enhance the nutritive properties of milk, in others to reduce putative milk allergens, and in yet others to produce substantial concentrations of high-value pharmaceutical proteins in milk such as factor VIII, factor IX, protein C and α_1-antitrypsin (in the latter case, up to 35 g l^{-1}). A recent advance whereby a gene construct is introduced into the egg's nucleus before fertilization has greatly improved the efficiency of preparing transgenic offspring. The horizon of diversification for the specialist stockkeeper has been expanded further with the transgenic modification of animal organs for xenotransplantation (Squinto, 1996). Although the latter example has little to do with food security, it serves to illustrate the extraordinary range of possibilities that now exists for the design, transfer and targeting of gene constructs tailored in precise ways to meet specific requirements (Houdebine, 1997).

Clearly, there is no shortage of options offered by biotechnology for application to the husbandry and production of livestock. The options outlined above are by no means exhaustive and they map on to those selected for the future direction of public agricultural research reported in the Strategic Plan of the USA (environment and natural resources; nutrition, food safety and health; added-value processes and products; economic and social issues; animal and plant systems; Lacy, 1995). Yet even in a country with such inherent wealth and resources, funding is unpredictable. One of the major factors is that animal biotechnology is characterized by a long lead time to application, due not only to the technological challenges and the infrastructure needed to train practitioners, but to the proper demands of animal welfare and the need to inform a sceptical public.

PUBLIC SCEPTICISM ABOUT EMERGING OPTIONS

Returning to the question about what makes biotechnology special, leading theorists in food systems have proposed that in the past there was a structural separation between agriculture and its downstream industries resulting from the dependence of the farmer on nature. Biotechnology has weakened this separation substantially because, as we have seen already, genetic engineering can be used to modify the biological process on the farm, to break down raw materials, or to engage in the production of high-value industrial outputs. It is also possible to use these new technologies in food processing (Fine *et al.*, 1996). Public concern about biotechnology as applied to food animals has alerted policy makers and scientists alike to the need for transparency, accountability and regulatory approval. In fact, regulatory approval has become almost as important as the scientific achievement itself if the applications of biotechnology are to become realized. Public perception, therefore, remains a major constraint in the adoption of many of the options so far considered.

The prospect that limits may be set regarding the use of biotechnology was apparent from the speech in 1989 of the former President of the European Union, Jacques Delors. He spoke then about human rights and the European Community and said

> the ethical dimension is once again coming to the fore, and we must step up the debate about these fundamental issues which concern the very essence of human life and society. On the basis of what scientists tell us about the laws of Nature, we must take responsibility and decide what action we want to take. For my part I would like to see the debate conducted in philosophical and ethical terms so that our understanding advances to keep pace with scientific progress.

This viewpoint has been represented in an extreme form by the

German Green Party which was active in seeking a biotechnology embargo driven by a deep-seated mistrust of genetic research. The Party has linked biotechnology with eugenics and social engineering because it has been seen as a disturbing extension of the horrendous events of the Third Reich. Recently, however, Germany has announced a substantial increase in the funding of biotechnology, underlining an intention to become the major force in Europe by the year 2000. This decision realigns Germany with the European Commission's White Paper on *Growth, Competitiveness and Employment* (1993) which stated that the confluence of classical and modern technologies enables the creation of new products and highly competitive processes in a large number of industrial and agricultural activities as well as in the health sector. The sequence of events serves to emphasize an ambivalence that, as with the outbreak of bovine encephalopathy in the UK, has done little to reassure a sceptical European public about the costs and benefits of the technologies in the agricultural sector.

ETHICAL ISSUES

Stephen Toulmin (1982), writing in 'How medicine saved the life of ethics', claimed that the advances of modern medicine rejuvenated the moribund discipline of ethics by the problems that it posed. Some would claim that modern agriculture and biotechnology have had a similar impact on bioethics. In utilitarian terms, many of the options of biotechnology identified above offer the potential to increase the quantity, quality and safety of food supplies by overcoming geographical and climatic obstacles. They could facilitate the rapid dissemination of high-quality stock with specific forms of disease resistance or superior genetic qualities compatible with sustainable practices. In these respects, they would seem to occupy the moral high ground if they are fair to animals and the environment, and are accessible to those farmers who have greatest need of their application.

To address public concern, the UK's Ministry of Agriculture Fisheries and Food commissioned two reports; *Ethics of Genetic Modification and Food Use, 1993* (the Polkinghorne Report, 1993) and *Ethical Implications of Emerging Technologies in the Breeding of Farm Animals, 1995* (the Banner Report, 1995). Common to the work of both Committees was the recognition of the need to safeguard individual and collective freedom of choice and action, the well-being of others, the aim to do no harm and social fairness. In addition to these traditional ethical principles of autonomy, beneficence, non-maleficence and justice, the issue of animal welfare was added.

THREE CASE STUDIES

Case study 1: breeding

The Banner Report focused on animal welfare and stressed that the dominant cultural view of animals has been anthropocentric. Humans, though part of nature, were taken to be superior to other animals and organisms as portrayed in Christian, Jewish and Moslem teachings. It restated that non-human animals are sentient fellow beings with an intrinsic value or inherent worthiness and are not mere instruments or production machines. The Report argued that the possible uses of some of the new reproductive technologies were intrinsically objectionable on the grounds that they failed to respect the dignity, natural characteristics and worth of animals. For example, it recommended that any procedure which aimed to reduce the sentience and responsiveness of pigs and thereby increase their efficiency of food conversion would be unacceptable for these reasons. It expressed concern that animals should not be treated as raw material 'upon which our ends and purposes can be imposed regardless of the ends and purposes which are natural for them'. It demanded that the cost–benefit analysis of any procedure should demonstrate that the technologies would bring substantial good if they were to be permissible, as in the case of non-therapeutic surgical interventions for embryo transfer and/or transgenesis. Latterly, there has been a further shift away from the anthropocentric position to a more 'eco- or bio-centric' way of thinking, and it is attitudes such as these that have driven the adoption of the precautionary principle – to grant approval for new procedures only when all preventative steps have been adopted.

Animal biotechnology has been the focus of suspicion because of a perceived absence of concern about risk, the lack of a regulatory track record and the apparent disrespect for the holistic nature of animals. Regrettably, examples exist where the recent record has been unhelpful and triumphalistic. Poultry have been bred by advanced (but conventional) selection procedures to reach slaughter weight in just 6 weeks, and this has been seen as symbolic of the power of genetic selection. The risk of bone pathologies has increased, and the strict monitoring of welfare has become crucial in intensive systems. Transgenic pigs expressing extra copies of GH were publicized widely because they showed a substantial increase in growth rate (Beltsville pig). Subsequently, it was found that accelerated growth was associated with unacceptable welfare problems including serious joint and reproductive abnormalities. (These pathological problems can be circumvented by the use of appropriate gene constructs which no longer result in the overproduction of GH.) The successful cloning of cattle by nuclear substitution has been reported, but it was apparent that substantial losses of embryos occurred during pregnancy and large-for-dates offspring resulted in dystocia and in an increase in delivery by Caesarean section. It has become essential to clarify by basic research the mechanisms that underlie these problems before application of these procedures can be seriously contemplated.

Case study 2: cloning

The recent report of cloning by the use of somatic nuclear substitution has provoked a plethora of reactions ranging from panic to euphoria and has attracted excessive media coverage (Hodges, 1997). For livestock, the technique has the potential for the production of large numbers of genetically selected and identical offspring of pre-determined sex, for the introduction of genes into the germline through the use of ES cells and, eventually, as relevant genes become known, for the production of disease-free animals. For example, the procedure has been proposed for the multiplication of replacement stock free from genes coding for prion proteins associated with transmissible encephalopathies.

The accepted wisdom prior to the Dolly experiment was that a donor cell originating from an adult mammalian species would already be irreversibly programmed to develop in a certain way. At least in the case of a cell taken from a primary culture of sheep mammary gland this has been proved not to be the case. Dolly was the result of fusion of an adult donor cell derived from a Finn Dorset ewe with an enucleated egg derived from a Scottish Blackface ewe. These breeds are phenotypically distinct, a tried and tested genetic device that nicely complements the all-important DNA fingerprint data in determining that Dolly was indeed derived from a Finn Dorset, and not a Scottish Blackface nucleus! After the removal of nuclei from donor eggs (Scottish Blackface ewes), a donor adult cell (Finn Dorset) was injected, fused and activated by passing an electrical current. Subsequently, the activated egg was cultured and then transferred into the uterus of a Scottish Blackface recipient ewe. The adult donor cell derived from a primary culture of mammary gland cells was forced into a so-called period of quiescence (deprived of serum during culture) before being transferred into the recipient egg. The success rate was low (1 out of 277 transfers) and, in parallel experiments with embryonic and fetal cell nuclei, a high incidence of embryonic and fetal mortality was also observed even during the late stages of gestation. This very high loss has not only given rise to ethical concerns about animal welfare but makes the process as it stands unacceptable on utilitarian grounds. Subsequent studies in mice and cattle have confirmed the possibility of deriving viable offspring after nuclear replacement with a somatic nucleus from an adult donor even without the imposition of quiescence for the reconstituted egg.

It is important to analyse the difficulties that lie ahead before any practical application of cloning can be seriously envisaged. First, a principal motivation behind mammalian cloning was to facilitate genetic engineering in the development of new transgenic animals of medical importance to humankind. The production of therapeutic pharmaceutical proteins in milk or blood remains a major objective, and the Roslin technique has provided a new way to achieve that end. Questions of animal welfare and the maintenance of genetic diversity in populations of elite stock need to be addressed if this approach is to be found acceptable by the public, scientists and welfarists alike.

Second, the production of 'nuclear' stock derived from banks of embryos of the desired and tested genotype, or the environmental application of cloning in terms of the preservation of species that are endangered or resistant to breeding in captivity will be seriously impeded if the efficiency of the procedure cannot be improved substantially. Progress has been made already, but if difficulties were to persist there will be little chance that cloning is just around the corner. Current advances include nuclear replacement prior to fertilization, which shows considerable promise.

Third, caution is needed when interpreting the claims for medical use. Expectations have been raised about new treatments for ageing, cancer and the regeneration of new cells and tissue from cloned embryos. This approach has been discussed as a way to circumvent tissue rejection providing help for cases including Alzheimer's, Parkinson's disease and leukaemia. We agree with the Group of Advisors on the Ethical Implications of Biotechnology in its Opinion requested by the European Commission (1997) who stated that the intention to produce genetically identical human embryos raises serious ethical questions about safety and the use of human beings as tools of biomedicine (instrumentalization). In this respect, the way ahead demands careful and thorough study, and a large amount of human experimentation would need to be contemplated to ensure that the procedure is safe, as indicated in a recent UK report by the Human Genetics Advisory Commission and Human Fertilization and Embryology Authority (1998).

A statement from the Royal Society of London (1997) has supported the call for the prohibition of human cloning while allowing the derivation of cell lines and tissues from embryos cultured for up to 14 days after fertilization. With respect to animals, cloning was considered acceptable on the grounds of biomedical advances, improved food production in the case of livestock, and preservation of endangered species. It was further recommended that cloned animals should be the subject of detailed welfare studies (including cognitive capabilities), their production should be subject to a prospective analysis of any danger resulting from a loss in variation to the gene pool, and strategies should be developed by relevant institutions for maintaining diversity. The Royal Society recognized the urgent need for a wider and well-informed public debate of the scientific, technical, ethical and moral issues to ensure the involvement of lay and specialist participants from different generations and groups.

Scientific advances of the type signalled by the birth of Dolly have highlighted the importance of the formulation of guidelines, if not legislation, to permit a full analysis of the implications of the new development. This will also facilitate the public debate of the deontological issues, i.e. clarification of concerns about the rightness of the procedure in terms of revealed knowledge, religious teaching and philosophical thought, and the extent to which the procedure would be a further step down a 'slippery slope' and would be socially unacceptable. The UK's working group of the Human Genetics Advisory Committee has recommended that the cloning of humans should be prohibited except in cases of mitochondrial disease and for research into how to grow a patient's own tissues to treat serious disease. In 1999 the UK government, however, asked for further information before taking a final decision.

Case study 3: food

The Polkinghorne Report on the ethics of genetic modification and food use included interviews with different groups of people who were concerned about the risks associated with the use of biotechnology for food production. Some considered that the procedures induced a 'moral taint' to the products which carried over to make their use in food ethically unacceptable. 'Moral taint' in this context consisted of a moral revulsion at the process such that no reasonable person would consider consuming the food produced by it. The concept was derived from another area; namely the use for transplant purposes of the organs of a prisoner who had died under torture. The concerns expressed about genetically engineered food came from: (i) the nature of the technology itself which involved the transfer of genes from one species to another; (ii) the 'unnaturalness' of the process, even though present livestock have been modified by humankind over countless generations and would bear only slight resemblance to their ancestors either in phenotype or genotype; and (iii) possible animal welfare implications in that suffering caused to animals in the production of food rendered such food ethically unacceptable, however minor the suffering. Certain groups regarded as 'tainted' any food produced by genetic modification if animals were used in any way in its production or testing.

The Polkinghorne Committee were unable to accept the existence of a 'moral taint' that would warrant a total ethical prohibition on the use of genetically modified food since, though there were ethical anxieties about aspects of genetic modification, a total ban would require the judgement that the whole process was morally defective. They recognized that the greatest concern originated from the prospect of food containing human genes and this was addressed in the following way.

When a transgenic animal or plant was described as containing 'human' genes, it reflected the fact that it contained copies of a gene originally obtained from that source. It was not technically possible to take a gene directly from a human cell and insert it into host material such as a sheep embryo or seedling. A series of *in vitro* cloning and amplification steps would have to be undertaken to prepare genetic material suitable for insertion into a new host. As a result, the original 'human' gene was 'diluted' to the extent that its concentration in the embryo was vanishingly small. Therefore, the chances of recovering the original human gene from the sheep embryo were much less than the chances of recovering a specific drop of water released into the oceans of the world. As a consequence what is present in each cell of a modified sheep is not a 'human gene' but a 'copy gene of human origin', a term adopted to distinguish human transgenes from human genes in a human being.

Three aspects of the technology of gene transfer were found to be not commonly understood. First, that the widely used technique for the isolation, characterization and transfer of a gene involved the use of cDNA and meant that the original donor material (mRNA) was destroyed in the preparation of the DNA before it was inserted into the host organism. This removed even the remotest possibility that the modified organism contained genetic material directly derived from the donor. Second, that while different species appeared so dissimilar, their genes were not necessarily so markedly different. Genes that coded for the production of important proteins common to different animal species have been found to be remarkably similar; e.g. the insulin and IGF-1 genes have virtually identical sequences in sheep, cattle and humans. Of course, when genes are so nearly identical, this raises the question of whether they can be considered peculiarly 'sheep' or 'human' in nature, and why should one choose to use a human gene if others are available to produce the desired effect. Third, genes used in the modification of food have been derived by copying a gene extracted from the donor organism. However, if the sequence of a gene was already known, it would be possible to construct the gene by chemical synthesis. If this process was to be adopted, it would raise the question of whether a synthetic 'human gene' should carry the ethical status of a gene derived from a human.

Different traditions take different views of the nature and status of organisms containing copy genes of human origin, though all major religions prohibit cannibalism. Consultations showed that for the Moslem, the transgene retained its human nature and remained subject to the dietary taboo. For the Jew, the host organism remained the dominant species and did not assume the character of the donor. A sheep containing a copy gene of human origin remained a sheep, because it was the animal regarded as a whole which determined its status. To remain kosher, the animal was to be both cloven-hoofed and cud-chewing, and if genetic modifications did not alter these characteristics, there was no objection. Other religious groups were found to fall somewhere between these two positions. Hindu and Sikh attitudes were closer to the Moslem view. The latter acknowledged, however, that with greater understanding of the processes involved, attitudes might change, particularly in the case of synthetic human genes.

Christian attitudes appeared to be closer to the Jewish position, although representatives felt it very likely that some individual members of their faiths would find the consumption of copy genes of human origin objectionable (even though it related to only one such gene amongst the 100,000 or so of the host organism).

Regarding the transfer of copy genes of animal origin to other animals, various religious taboos prevented the consumption of certain types of meat. In particular, the consumption of pork was forbidden in Islam and Judaism, whereas Sikhism forbad the eating of beef. Hindus were usually vegetarian, but those that did eat meat would not eat beef. In general, the attitude of these groups to the consumption of food containing copy genes originating from 'proscribed species' corresponded to their position on the consumption of copy genes of human origin. Moslems, Sikhs and Hindus found such a practice objectionable, whereas Jews did not. Here again, attitudes were determined by the view taken of the introduced gene and whether or not it retained the nature and status of its origin.

In addition to the views of the religious groups, the genetic modification of food animals also raises a more general concern that modification may be used to secure economic advantage without proper regard for animal welfare. Such emphasis on utilitarianism may in part have been fuelled by the early Beltsville experiment in the USA referred to above involving the modification of pigs with the human GH gene.

Whereas no controls can totally prevent all unforeseen developments, legislation formulated along the lines of the UK Animals (Scientific Procedures) Act of 1986 would have demanded termination of any experiment which raised welfare concerns of this type. A case for additional welfare controls in the UK, specifically in relation to genetic modification programmes involving animals, was not recommended in the Polkinghorne Report in view of existing legislation and guidelines. Neither was there considered to be a case for special labelling rules to deal with animal welfare issues in relation to genetically modified animals, beyond any systems devised to meet more general welfare concerns.

Transfer of copy genes of animal origin to plants raised the issue of the attitude of vegetarians, and particularly vegans who avoid the consumption of all animal products. There was considerable diversity of dietary practice within vegetarianism. Those vegetarians who objected to meat-eating solely because of concern over animal husbandry and slaughter did not object to food derived from plants containing copy genes of animal origin or bacteria. Chymosin enzymes used in cheese production, when produced by genetically modified yeast or bacteria, represented an alternative to the traditional source of enzyme (rennet) obtained from calves stomachs. They were acceptable to vegetarians because the enzyme was thus calf-like although it was obtained without any animal involvement. Some individuals, however, found the presence of even a single copy gene of animal origin in a plant species to be unacceptable.

The Polkinghorne Report noted the absence of any overriding ethical objection which would require the absolute prohibition of the use of organisms containing genes of human (or non-human) origin provided the necessary safety assessment had been fulfilled. Whereas all consumers and religious groups consulted asked for some form of labelling to enable ethically necessary, informed choice, there was a good deal of reluctance on the part of industry to accept labelling because it stigmatized the product, raised many practical problems and was viewed negatively, as it had been with food additives and food irradiation. To meet the ethical concerns of those groups or individuals who objected to the consumption of food containing copies of genes of human origin, or from species which are the subject of dietary restrictions for their religion, or because of the views held by vegetarians, it was recommended that products should be labelled to allow consumers to exercise choice. A later opinion sought from the Group of Advisors on the Ethical Implications of Biotechnology by the European Commission (1995) concluded similarly that modern biotechnology as a technique cannot be regarded in itself as ethical or non-ethical. Labelling of food derived from modern biotechnology processes was considered appropriate when it caused a substantial change in composition, nutritional value or the use for which the food was intended. The European Parliament has now adopted a resolution calling for GM products to be labelled as such and sold separately from non-modified products, a political constraint that raises the prospect of confrontation with the World Trade Organization (WTO) which allows exclusions only on strictly scientific grounds. The Group of Advisors recommended that new ways should be developed to ensure that the public gained an objective and correct picture about the changes that occur in foods derived from the use of these technologies.

GLOBAL ETHICS

A criticism frequently levelled is that the affluent countries are preoccupied with the ethical implications of biotechnology while one-quarter of the world's population remains disadvantaged. Expressed in another way, the more developed countries consider that bioethics is a luxury the less developed cannot yet afford. However, ethics and values are distinct elements of any pluralistic civilization and of our cultural identities, and this leads us to consider where scientists get their ethical standards.

The behaviour of scientists in pursuit of new knowledge involves the collection of original information and the formulation of hypotheses tested for repeatability by methodology that has been evaluated rigorously and which is readily transferable to other investigators. A recent analysis of the behaviour of business leaders in the UK revealed three highly influential stereotypes insofar as their approach to the world of business was formulated (Boswell and Peters, 1997). For scientists, these stereotypes can be translated into first, those who are allergic to public policy thinking and social ideas because they are so preoccupied with running their experiments that they have little time for anything else; second, those who are concerned to follow the 'top-down' policy guidelines given by their superiors whether Directors of Research Councils or Heads of Funding Agencies; and third, those who are influenced by the latest pattern of ideas because they will attract others and give a sense of security in numbers and the prospect of funding.

At the bench level, the activity of scientists comes as either a projection or an extension of a single, potent individual whose original insights open new avenues of thought, or from their corporate dedication to an organization that seeks to solve an intractable problem of great social significance, or from the close interactions with the business world where the sights are set on products or processes, and on the profit that may flow from science and technology. A voluntary code of biotechnology ethics (or standards as some would insist) would undoubtedly stress truth in publication, full revelation of experimental findings even where conflicts exist, personal integrity in the pursuit of career or company prospects, and the rejection of narrowly defined technical–economic ideas about the management of science. For the western business leaders, invoking concepts of value or virtue familiar in the Christian, Greek or Hebraic traditions proved particularly attractive to some who were religiously minded, though the wider issues of co-operation and competition were avoided in favour of parochial issues such as an organization's hierarchy or mission (Boswell and Peters, 1997). Such is the public pervasiveness of science and technology and its international influence, that the demand for accountability and transparency in modern society leads scientists into the complex area of society's values and aspirations – albeit 'to do no harm'.

Clearly, biotechnology carries disproportionate benefits for some societies, and disproportionate costs for others, and to ask who gains and who loses is to ask ethical questions. Two examples illustrate the nature of the debate. First, less developed countries have raised the issue of the preservation of biological diversity and the desirability that benefits from valuable genetic material should accrue to the country of origin. Examples exist of partnerships between less and more developed countries where this problem has been addressed, but the adoption of an appropriate cost–benefit analysis has now become a priority for future collaborations. Second, the benefits derived from the use of bST to stimulate lactation in dairy cattle have failed to be realized in many less developed countries where the need is greatest because of the constraints imposed in more developed economies. In Europe, the decision to ban its application derived from socio-economic and political considerations rather than scientific or safety issues and, in consequence, severely restricted the development of the product.

A further issue where less developed countries are disadvantaged concerns the accessibility to new knowledge. Science operates with an authority and respect gained by being open with the knowledge it acquires, sharing it widely and being exposed to critical peer review. The movement of private interests into primary research, however, has brought pressures on scientists to patent their findings and consequently be less open with their results and ideas. Patenting was intended originally to discourage trade secrets and to encourage open disclosure of technical information while protecting the claimed invention. The paradigm shift came in 1985 when the USA allowed the extension of patenting of microbes to include genetically engineered plants, seeds and plant tissues. It was extended in 1987 to 'multicellular organisms, including animals' and in 1988 to the grant of patent on the Harvard oncomouse. The oncomouse patent was first filed with the European Patent Office in 1985, and the debate about the validity of this patent still rages. It is unsurprising that the USA currently holds 65% of biotechnology patents against just 15% held by EU countries.

Financial return, however, is not restricted to patented inventions; enormous wealth can still be generated by open publication. Monoclonal antibodies produced by a non-patented procedure resulted in estimated global sales of $2 billion per annum in 1993, which could rise by 5–20% per annum according to a recent European Commission White Paper (1993). Clearly, absence of patent protection does not preclude public good benefits provided the idea is a really good one! As far as food security is concerned, lack of access to intellectual property and the products of biotechnology by the less developed economies will continue to present a serious constraint unless imaginative partnership schemes can be secured to promote technology transfer and dissemination.

CONCLUSIONS

What of the future? The future is difficult to predict, but as Nobel Laureate and inventor of holography, Dennis Gabor, wrote 'futures can be invented'. For food security do we invent a future that includes biotechnology to enlarge the options for enhanced food security? Another Nobel Laureate and Advisor to the President of the US, Baruch Blumberg, writing in the Cambridge Review (1996) presented a futuristic view which we have adapted for the present purpose. He imagined one Carl Jenson, living in 2020, who is on a walking holiday with his family. In an idle moment, Carl recalls that the Human Genome Project has been completed; individualized preventive medicine, individualized nutrition and individualized allergen avoidance have become a reality. He has recently bought a personalized chip that carries all the necessary information about his genome. After a long day in the mountains, he sits around the campfire at Granite Lake, with his carton of long-lasting milk engineered to provide a composition redolent with high energy, low fat and devoid of allergens. He fries the beautiful fish he caught earlier, genetically engineered to encourage growth in a high altitude environment and a short growing season, and to provide high-quality protein compatible with his personalized nutritive demands. He examines the latest read-out from his chip and admits to all within earshot 'Well, I've learned to get along with what I have'!

The uptake of new ideas depends in the first instance on the successful demonstration of efficacy and safety in humans and animals. In the present context, they are most likely to be publicly acceptable if they lead to more sustainable production techniques and improved conservation of the environment including reduced land degradation. The increasing importance attached to the public acceptability of scientific advances is reflected in the investment by the US Department of Agriculture (USDA) which has awarded $6.4 million in research grants since 1992 (~ 1% of USDA spend on biotechnology research) to study the ecological risks of genetically engineered organisms developed for use in agriculture.

We have seen that there is a historical legacy in Europe which has contributed to a lack of confidence about public acceptability, and this has been reflected in the relatively weak activity in the entrepreneurial and venture capital environment, a prominent feature in the USA. In addition, the adoption of the precautionary principle in relation to the introduction of biotechnological options for livestock (as well as plants and microbes) has imposed a legitimate and inescapable constraint on the incorporation of these technologies for food production. Furthermore, if any of the options outlined above prove successful, they are unlikely to completely displace the more traditional forms of animal production and husbandry. The Technology Foresight exercise in the UK across 15 sectors of the economy concluded that genetic and biomolecular engineering, sensors and sensory technology are key generic priority areas for the future, with environmentally sustainable technology as an intermediate area. They were firmly placed in the most important quartile of an analysis that included the criteria of attractiveness and feasibility. Health and lifestyle appeared as the most attractive, yet least feasible key generic priority. Food production failed to make the final chart!

If food security is not the flavour of foresight, the doomsday scenario of the Reverend Thomas Malthus of Jesus College, Cambridge, written nearly 200 years ago may yet come to haunt us. Many observers believe that the implications of the Malthusian model, which predicted a population growth rate that would outstrip food supply, have been averted precisely because of the technological advances since that time.

SUMMARY

Today's world food demands could not have been met with the farming methods of the 1940s, and there will be little prospect of responding to the needs of the next century without advances in the methods of the

1980s. One prospect is the adoption of biotechnology to increase the range of options available to food producers and processors. We examine some of the emerging options which are beginning to have an impact on genetic selection, animal health, production and efficiency, animal nutrition and reproduction, and niche markets for specialized animal products.

People replace plant-derived foods with animal products as material standards of living rise. The options offered by biotechnology, whether for plant or animal products, have revealed constraints and limits which derive partly from ethical concerns about the modification and control of gene function, and partly from a deep-seated distrust of powerful alliances between science and industry, particularly in more developed economies and especially with the growing involvement of multinational companies. The complex reasons for, and the influence of, these constraints and limits need to be understood if options are to be turned into reality. The processes of regulatory approval and gaining public acceptability have now become almost as important as scientific discovery.

Part of this paper is taken from 'Animals and the human food chain', Heap (1998).

REFERENCES

Ali, S., Hall, J., Soole, K.L., Fontes, C.M.G.A., Hazlewood, G.P., Hirst, B.H. and Gilbert, H.J. (1995) Targeted expression of microbial cellulases in transgenic animals. In: Petersen, S.B., Svensson, B. and Pedersen, S. (eds), *Progress in Biotechnology 10: Carbohydrate Bioengineering. Proceedings of an International Conference,* Denmark. Elsevier, Amsterdam, pp. 270–293.

Andersson, L., Archibald, A.L., Ashburner, M., Audun, S., Barendse, W., Bitgood, J., Bottema, C., Broad, T., Brown, S., Burt, D., Charlier, C., Copeland, N., Davis, S., Davisson, M., Edwards, J., Eggen, A., Elgar, G., Eppig, J.T., Franklin, I., Grewe, P., Gill III, T., Graves, J.A.M., Hawken, R., Hetzel, J., Hilyard, A., Jacob, H., Jaswinska, L., Jenkins, N., Kunz, H., Levan, G., Lie, O., Lyons, L., Maccarone, P., Mellersh, C., Montgomery, G., Moore, S., Moran, C., Morizot, D., Neff, M., Nicholas, F., O'Brien, S., Parsons, Y., Peters, J., Postlethwait, J., Raymond, M., Rothschild, M., Schook, L., Sugimoto, Y., Szpirer, C., Tate, M., Taylor, J., VandeBerg, J., Wakefield, M., Wienberg, J. and Womack, J. (1996) Comparative genome organization of vertebrates: The First International Workshop on Comparative Genome Organization. *Mammalian Genome* 7, 717–734.

Armstrong, D.G. and Gilbert, H.J. (1991) The application of biotechnology for future livestock production. In: Tsuda, T., Saaki, Y. and Kawashima, R. (eds), *Physiological Aspects of Digestion and Metabolism in Ruminants: Proceedings of the Seventh International Symposium on Ruminant Physiology.* Academic Press, San Diego, pp. 737–761.

Banner, M. (Chairman) (1995) *Report of the Committee to Consider the Ethical Implications of Emerging Technologies in the Breeding of Farm Animals.* Ministry of Agriculture, Fisheries and Food, HMSO, London.

Blumberg, B.S. (1996) Medical research for the next Millennium. *The Cambridge Review* 117, 3–8.

Boswell, J. and Peters, J. (1997) *Capitalism in Contention.* Cambridge University Press, Cambridge.

The Brundtland Report (1987) *The Brundtland Report – Our Common Future.* Report of the World Commission on Environment and Development, Oxford University Press, Oxford.

Burt, D.W., Bumstead, N., Bitgood, J.J., Ponce de Leon, F.A. and Crittenden, L.B. (1995) Chicken genome mapping: a new era in avian genetics. *Trends in Genetics* 11, 190–194.

Damak, S., Jay, N.P., Barrell, G.K. and Bullock, D.W. (1996a) Targeting gene expression to the wool follicle in transgenic sheep. *Bio/Technology* 14, 181–184.

Damak, S., Su, H-Y., Jay, N.P. and Bullock, D.W. (1996b) Improved wool production in transgenic sheep expressing insulin-like growth factor I. *Bio/Technology* 14, 185–188.

Devlin, R.H. (1997) Transgenic salmonids. In: Houdebine, L.M. (ed.), *Transgenic Animals – Generation and Use.* Harwood Academic, Amsterdam, pp. 105–117.

Donnelly, J.J., Ulmer, J.B., Shiver, J.W. and Liu, M.A. (1997) DNA vaccines. *Annual Reviews of Immunology* 15, 617–648.

Draghia-Akli, R., Li, X. and Schwartz, R.J. (1997) Enhanced growth by ectopic expression of growth hormone releasing hormone using an injectable myogenic vector. *Nature Biotechnology* 15, 1285–1289.

Dyson, T. (1996) *Population and Food – Global Trends and Future Prospects.* Global Environmental Change Programme, Routledge, London.

European Commission White Paper (1993) *Growth, Competitiveness and Employment – the Challenges and Ways Forward into the 21st Century.* ECSC – EC – EAEC, Brussels Luxembourg, 1994.

European Commission (1995) Ethical Aspects of the Labelling of Foods derived from Modern Biotechnology: opinion requested by the European Commission. *Opinion of the Group of Advisers on the Ethical Implications of Biotechnology to the European Commission, 1995.* Brussels, Belgium.

European Commission (1997) Ethical Aspects of Cloning: opinion requested by the European Commission. *Opinion of the Group of Advisers on the Ethical Implications of Biotechnology to the European Commission, 1997.* Brussels, Belgium.

Fine, B., Heasman, M. and Wright, J. (1996) *Consumption in the Age of Affluence – the World of Food.* Routledge, London.

Georges, M. (1997) Recent progress in mammalian genomics and its implication for the selection of candidate transgenes in livestock species. In: Houdebine, L.M. (ed.), *Transgenic Animals – Generation and Use.* Harwood Academic, Amsterdam, pp. 519–524.

Haley, C.S. (1995) Livestock QTLs – bringing home the bacon? *Trends in Genetics* 11, 488–492.

Hammond, K. (1993) Why conserve animal genetic resources. *Diversity* 9, 30–33.

Heap, R.B. (1998) Animals and the human food chain. In: Waterlow, J.C., Armstrong, D.G., Fowden, L. and Riley, R. (eds), *Proceedings of the Rank Prize Funds Conference on Feeding a World Population of More Than Eight Billion People: a Challenge to Science, December 1996.* Oxford University Press, New York, pp. 232–245.

Heap, R.B. and Moor, R.M. (1995) Reproductive technologies in farm animals: ethical issues. In: Mepham, T.B., Tucker, G.A. and Wiseman, I. (eds), *Issues in Agricultural Bioethics*. Nottingham University Press, Nottingham, pp. 247–368.

Hillman, J. (Chairman) (1995) *Progress through Partnership: Technology Foresight – Agriculture, Natural Resources and Environment*. Office of Science and Technology, London.

Hodges, J. (1990) Animal genetic resources. *Impact of Science on Society* 40, 143–154.

Hodges, J. (1997) Cloning sheep – cloning people? (Editorial). *Livestock Production Science* 49, 63–68.

Houdebine, L.M. (ed.) (1997) *Transgenic Animals – Generation and Use*. Harwood Academic, Amsterdam.

Human Genetics Advisory Commission and Human Fertilization and Embryology Authority, UK. (1998) *Cloning Issues in Reproduction, Science and Medicine*. Report from The Human Genetics Advisory Commission and The Human Fertilization and Embryology Authority.

Jun Du, S., Gong, Z., Fletcher, G.L., Shears, M.A., King, M.J., Idler, D.R. and Hew, C.L. (1992) Growth enhancement in transgenic atlantic salmon by the use of an 'all fish' chimeric growth hormone gene construct. *Bio/Technology* 10, 176–181.

Kline, D. (1990) Agricultural bioethics and the control of science. In: Gendel, S.M., Kline, A.D., Warren, A.D. and Yates, F. (eds), *Agricultural Bioethics – Implications of Agricultural Biotechnology*. Iowa State University Press, Ames, Iowa, pp. xi–xx.

Lacy, W.B. (1995) Socio-economic context and policy strategies for US public agricultural sciences. *Science and Public Policy* 22, 239–247.

Malthus, T. (1798) *An Essay on the Principle of Population and a Summary View of the Principle of Population*, Flew, A. (ed. 1970). Re-issued Penguin Classics, London, 1985.

McMichael, A.J. (1993) *Planetary Overload – Global Environmental Change and the Health of the Human Species*. (Canto edition 1995.) Cambridge University Press, Cambridge.

Mercier, J.C. and Vilotte, J.-L. (1997) The modification of milk protein composition through transgenesis: progress and problems. In: Houdebine, L.M. (ed.), *Transgenic Animals – Generation and Use*. Harwood Academic, Amsterdam, pp. 473–482.

National Bioethics Advisory Commission (NBAC) (1997) *Cloning Human Beings*. Report by American NBAC, Rockville, Maryland.

Polkinghorne, J.C. (Chairman) (1993) *Report of the Committee on the Ethics of Genetic Modification and Food Use*. Ministry of Agriculture, Fisheries and Food, HMSO, London.

Rifkin, J. (1992) *Beyond Beef – the Rise and Fall of the Cattle Culture*. Dutton Books – Penguin Books, USA.

Royal Society of London (1997) *Whither Cloning*. Report of the Royal Society, London.

Sen, A. (1995) Food, economics, and entitlements. In: Dreze, J., Sen, A. and Hussain, A. (eds), *The Political Economy of Hunger – Selected Essays*. Clarendon Press, Oxford, pp. 50–68.

Spedding, C.R.W. (1984) New horizons in animal production. In: Nestel, B.L. (ed.), *World Animal Science A2 – Development of Animal Production Systems*. Elsevier, Amsterdam, pp. 399–418.

Squinto, S.P. (1996) Xenogeneic organ transplantation. *Current Opinions in Bio-technology* 7, 641–645.

Toulmin, S. (1982) How medicine saved the life of ethics. *Perspectives in Biology and Medicine* 25, 736–750.

Wilmut, I., Schnieke, A.E., McWhir, K.J., Kind, A.J. and Campbell, K.H.S. (1997) Viable offspring derived from fetal and adult mammalian cells. *Nature* 385, 810–813.

Animal Welfare and Use

3

Donald M. Bruce and Ann Bruce

The Society, Religion and Technology Project, Church of Scotland, Edinburgh, UK

INTRODUCTION

Animal welfare is considered in relation to ethical and social values and norms, in its widest context, taking into account not just physical harm but also other aspects of intervention. The way in which we treat animals is seen as a reflection of how we view our relationship not only with them but also with nature in general.

Five models are described to illustrate a range of views towards nature – ownership, maintenance engineer, Christian stewardship, partnership and worship. They represent a spectrum of intervention. The two poles are the ownership model, where nature is there purely for human use and disposal, and the worship model which implies the severest limitations to what we may do to nature. In between lie more practical viewpoints. Partnership with nature presents a less restrictive view of intervention than the worship model, and the maintenance engineer is a moderated, pragmatic version of ownership. Between these two comes a Christian model of relationship and stewardship and, taking into account the concept of God who is seen to exist outside nature, and the balance of intervention and respect which this leads to. While the Christian model is the one most familiar to the authors, it is likely to have resonance with other fellow travellers.

These models are to some extent artificial in their delineations. They are intended to show some of the different ways in which it is possible to approach animal use and welfare. While they do not express particular world views, they can help us to clarify where we stand, why we think what we think, and what underlying motives and values drive our actions, and, if necessary, to change them.

These models are then used as a framework against which to evaluate animal welfare as expressed in a selection of current developments in animal breeding, genetics and cloning. These include marker-assisted selection, genetic engineering in principle and practice, and three examples of its application – pharmaceuticals in milk, xenotransplantation and disease models – and, finally, several aspects of animal cloning. These cases reveal how the limits of acceptability vary from model to model, depending on the example. Animal production takes place in a social context, and this is briefly considered in relation to scientific and economic forces and public accountability. In the western industrial context this is tending to push towards the problems of the 'ownership' model of relating to animals. It is therefore up to us to find a way of working in this context to achieve the quality of life in the 21st century that we would wish to aim for.

ATTITUDES TO ANIMALS

There is no simple definition of animal welfare. Duncan and Fraser (1997) have pointed out that the term did not arise as a scientific concept but as a reflection of our value system, to express our concern for the appropriate treatment of animals. They refer to three types of definition, based on:

- the experiences of the animals, such as pain or pleasure;
- normal biological functioning of the animal;
- the nature of the species and ensuring the animal can exhibit its full range of behaviour.

It is thus not just a question of 'not doing harm' to an animal, but rather it involves evaluating a complex array of different aspects. It also reflects on the quality of human life in the 21st century. What does treating animals in a particular way imply about us, and do to us, as individuals and as societies? There is a paradox. At the same time that our technical ability to exploit farm animals is expanding rapidly into entirely new fields, such as cloning and gene replacement, in many countries of the north increasing concern is being expressed about the limits to which we may use animals instrumentally, as means towards ends. This chapter examines these more fundamental issues about what is appropriate intervention in the animal kingdom, beyond the question of degrees of harm.

To ask what may be permitted towards animals means first establishing what is our relationship to them. In order to help us understand our relationship to animals, it is important to place this in the context of a much wider question, of our basic attitude towards nature as a whole. The very language which people use for the natural world which we find around us – expressions such as the environment, God's creation, the Earth, natural resources, biosphere, Mother Nature, forces of nature, and so on – is itself symptomatic of underlying value judgements. These valuations have profound effects on how we think 'nature' should be treated. Our view of how we treat animals stems from how we view nature, which is often so much part of a culture as to be invisible. Appreciating the underlying reasons can help clarify our understanding and thus our actions.

At different times and in different cultures, particular views have predominated. At other times, as with the current complex and fragmented philosophical milieu in Europe and North America, many different views exist side by side, with no one view holding sway. Amongst these one can identify four broad types or categories, which we have named (Bruce, 1993):

1. Ownership
2. Worship
3. Partnership
4. Maintenance engineer.

We discuss these four types and also propose a fifth model, which initially is derived from a Christian understanding of notions of creation, stewardship and relationship, but has a much wider applicability.

The categories represent a spectrum of permissible interventions in nature, defined by a model of different notions of relationship to nature. The choice to use 'relationship' as the measure is taken, in our case, from a Christian understanding of the world in which the relational element is crucial, but it is something with a much broader application in many other belief and philosophical systems. While the terms we use have been coined in the context of an industrialized society, the models of the natural world which underlie each of them are more universally applicable.

From these types come corresponding understandings of what is regarded as the proper role of humanity towards nature, and thus towards animals. These are stereotypes, identified to try to make clearer what the range of choices is, and what are the consequences of those choices. These are like grid references of four major topographical features on a map, which may help the reader recognize his or her own position, and navigate accordingly. Something similar is also being explored by Visser and Verhoog (1999), although without the fifth model.

Model 1: ownership of nature by humans

At one extreme, attitude towards nature can be expressed by the paradigm of ownership. This is the idea that nature is there purely for human use and disposal. It is ours to use and exploit exactly as we wish. We, as humanity, just 'found' it, as it were. Since no one seems to own it, it becomes a case of 'finders keepers.' Having staked our claim as its owners, we can do what we like with it. There is thus no restraint on its use. Whatever way we find to use nature, we use it. The dominant motive is that whatever is good for humans is good. It is the model of complete anthropocentricity, making an absolute disjunction between humans as the subject and nature as pure object. On this value system, if there is money to be made or something in nature to be exploited, then why not do it?

Another way of putting it is the sense of mastery. Here the idea is one of overcoming an opponent. If our experience of nature leads us to see it as a threat, then it has to be conquered. Its unruly forces must be overcome and held down, and become instead resources for humanity to harness. There is an element of pride. Humankind at the top of the evolutionary tree must prove its power and overlordship. This is the biblical creation story notion of 'subdue the earth' cut loose from its counterbalancing notion of 'care' (*Genesis* 1:28 and 2:15).

At the heart of this type is the notion of nature as 'it', an external object to which I am the subject – the creation of the classic Cartesian duality between observer and observed. This is in turn related to an Enlightenment scientific ideal, of the human observer dispassionately examining the object, in this case animals, and then using them accordingly. In this respect, ownership is a *lack* of relationship. This view is uncompromisingly instrumental towards animals. If we are absolute owners, it follows that animals do not have any status in themselves morally. Their value is only that which humans can make of them. Outside this, they have no value. In such a model, selective breeding has no limits, and genetic modification of animals for human benefit or convenience presents no questions whatsoever. It is pointless to query the xenotransplantation, say, of a modified pig's heart into a human. Since there is a human benefit to be gained, there is no question to be answered about the use of the animal, or its welfare. We would simply ask what would maximize the efficiency of its use. Questions of whether salmon should be genetically modified for faster growth rate or breeding stock cloned to produce a consistent product elicit the simple response 'why not?'

Model 2: worship of nature

At the other end of the spectrum would be a view that we should do almost nothing with nature. There are various ways this could be expressed, but perhaps the best illustration is that of worship as the opposite pole to ownership. Nature is divine, either because it is god in the sense of 'Mother Nature' or it is seen as an expression of god in a very direct sense – what might be termed panentheism or god in everything. This has two aspects. To the extent that nature is divine or holy, it is due our most profound respect. This implies severe limitations on what we may do to alter it. The ordering and relationships throughout the natural world are in themselves part of the divine wisdom that we are not given to tamper with. At one level, we should not disturb part of the divine pattern as a mere part of that pattern ourselves. At another, do not do to cattle what we would not do to humans, because we are both equally divine.

Secondly, to the extent that nature is divine, the model is also eco-centric. Humans are merely one part of a complex creation, instead of being at the centre of it, as in the ownership model. The stress is on similarity and congruency, based on things shared in common, instead of emphasizing the differences. In direct contrast to the subject–object connection of ownership, here it is subject–subject. This is also consonant with a secular environmental viewpoint from within what has become known as the 'deep ecology' tradition, in reaction to the excesses of exploitation under an ownership model. Frequent lament is made of the 'desacralizing' of nature, the removal of the sense of sacredness of the earth and its creatures, its seasons and its varied phenomena. The blame is ascribed variously to western Christianity, modern science emerging out of the Reformation and, perhaps most accurately, to the European Enlightenment.

The end point is much the same. In stressing either the divinity or the intrinsic relatedness of all beings, a large question mark is placed over how far it is acceptable to manipulate one species for the benefit of another. In the limit, this paradigm would allow no use of animals at all. In practice, very few belief systems actually worship all animals as divine to the point that it is considered wrong ever to use them in any way; however, in both the Jains and the extremities of the UK animal rights movement, one sees something close to a complete restraint on use.

Extremes and realities

The above two models are knowingly presented as stereotypes, to make the point that there are two opposing tendencies, clearly recognized by their extremes, but not normally expressed as such. There is also a sense that culturally one leans towards the West, expressing a materialistic and scientific outlook, the other towards the East and the mystic frame of mind, or at least looking to sources of value other than the instrumental and utilitarian. Again one would not wish to overemphasize the point, because these are not intended to express any particular world view, so much as to explore what attitudes underlie the extremes of unrestrained exploitation or radical non-intervention in nature in general and the animal kingdom in particular.

In practice, the vast majority of both religious and secular perspectives draw back from the view that literally anything is permissible to do in nature or that literally nothing should be done to alter it. However, most show a tendency towards either the one or the other. These could be described as a 'partnership' with nature, as a moderated view of the worship model, and the 'maintenance engineer' as a more pragmatic version of ownership.

Model 3: partnership with nature – a reluctant use

Partnership with nature removes some of the prohibitions on intervention by which we defined the worship model. Nature is no longer so 'divine' that it is untouchable, but a deep sense of respect remains. We may act, but not too much. The changes we make must not upset the sense of an overall balance. The relationship is still that of one member of nature to another, of equals, not of higher or lower. The expression of this broad perspective can be seen in the environmental movement in industrialized countries. Its roots lie, amongst others, in the European Romantic tradition, reacting against the Enlightenment's mastery over nature, and in America the writings of John Muir, the emerging of the wilderness movement and the founding of the Sierra Club. Nature is seen as a source of inspiration rather than of worship, of worth in itself, and of fragile beauty which must not to be diminished either by callousness or carelessness on the part of humans.

Amongst non-industrial, indigenous cultures, the native American practices are often cited, particularly in their attitudes to animals. The sense is that of a cooperative reciprocal relationship of equals. We are allowed to use them and eat them, but we must remember that we are part of nature too, and that our life and that of all things around us are as one. In western expressions of this model, animals are seen to require moral consideration and to have intrinsic value. The controversial notion of animal rights would fall within this category. The problem with this model is where and why to draw lines, without simply following personal preferences. Regan argues that animals, being 'subjects of a life' have duties owed them almost the same as those owed to humans (Regan, 1988), which goes closer towards the worship category, in ruling out hunting or rearing animals for food production and genetic engineering on principle. From a very different angle, Singer objects to the species supremacy assumed by humans and adopts a utilitarian argument for equal moral duties to animals on the basis of a common bond of sentience (Singer, 1990). For the purposes of this analysis, however, we would define the partnership model as the reluctant use of animals, under close limits – a 'No, unless' policy. Under this definition, genetic modification is questionable and intensive industrial animal production systems are strongly criticized as inappropriate.

Model 4: the maintenance engineer

If partnership is 'No, unless', the moderated version from the ownership side is 'Yes, provided.' What we call the maintenance engineer looks after nature, not altruistically but because it is in our best interests to do so. This model is inherently pragmatic. The former UK prime minister Mrs Thatcher famously referred to humans having a 'full repairing lease' on the planet (Thatcher, 1990). Visser and Verhoog (1999) refer to the 'rentmeester' to describe a very similar category. It is our responsibility to look after nature, often referred to in terms of the Christian notion of stewardship, as discussed below, but it is a fully anthropocentric sense of what the steward is responsible to. We are responsible to ourselves and our future generations, with humans still very much at the centre of this universe.

Under this model, animals may be used widely for all sorts of purposes but there are also limits. Genetically modified sheep are valuable for the pharmaceuticals in their milk, therefore we must look after them. Bad husbandry is bad business. Using bovine somatotropin (bST) is acceptable provided that we do not push up milk yields under poor husbandry conditions and so cause welfare problems (Wilson, 1998). It is not so much that other living beings have rights, but they are also sentient creatures, towards whom there are limits of what we may do. Animal welfare considerations will moderate the ways animals may be used. It would not, however, extend to Singer's argument for *equal* moral duties to animals, on the basis of sentience for example. We look after animals primarily because they are valuable to us.

Some practices in traditional agriculture may also impinge greatly on the animal. A *de facto* utilitarian attitude towards animals is not the sole prerogative of western patterns of thought.

Summary of the four categories

The above scheme asserts that a spectrum of levels of permissible intervention in animals can be seen, on the basis of underlying attitudes to nature. Within each of the four categories, a number of different philosophical or religious approaches and traditions might motivate the justification of a broadly similar end, with greater or lesser conformity. Inevitably there are some awkward fits. Religious animal sacrifice lies largely beyond the scope of this book, but is an example of a more complex contextual situation. Although it might fall within the maintenance engineer category, in that the use of the animal is instrumental, it is done for a different reason than human utility alone, because something other than humans is being sacrificed to. If sacrifice is made to appease or persuade otherwise capricious or malevolent forces of nature, perhaps to ensure fertility or a good harvest, distinctions are made among different aspects of nature. Even here there is a utilitarian element, the few being sacrificed for the good of the many.

Each of the four categories, as described, could be criticized as the basis for practices towards farm animals. Ownership and worship are extremes, and thus have the severe difficulty of the denial of the significance of ethical categories other than the one regarded as paramount. As we have defined it, ownership accords no significance to the animal and no restraint towards it. The worship model can raise conflicts over denials of human welfare, and has the more pragmatic problem that strict non-intervention by humans is almost impossible, as well as a logical problem of where the line is drawn. In its moderated utility, the maintenance engineer still misses the sense of intrinsic worth, whereas the partnership model only grudgingly admits any utility at all. This seems to point to the need for a fifth category intermediate between the latter two, and also to establish a due basis for the necessary decisions to be made among competing factors.

Model 5: a Christian or intermediate model

In the argument given, the four categories have assumed an essentially closed system for the universe. By definition, this misses the aspect of religious traditions where God is seen to exist 'outside' nature. In particular, Judaism, Christianity and Islam bring this external perspective to the spectrum of possible interventions in nature. The following argument seeks to set out a Christian framework for animal intervention, but drawing attention to features that would be recognized as held in common by many who did not ascribe to the Christian faith, but may none the less be fellow travellers with some of its presuppositions.

In this understanding, the universe is created by God as something external to, but continually recreated and sustained by, God. The world around us is thus creation, not merely 'nature', and it belongs to God, not human beings. (Note that the use of the word 'creation' in this context simply means that all things in the universe owe their origin and sustaining ultimately to God. It should not be read to imply a particular interpretation of some that the universe was created in six 24-h days, but it could equally encompass an evolutionary model over a very long time frame, with God as its author.)

God's ways are revealed not only by human wisdom but also by direct revelation found in the teachings of Christ and the writings of the biblical records. These give a picture in which the creation, including animals, has intrinsic value, because God created it. Animals exist first of all to God, before any considerations of their value and use to humans. This view is neither ecocentric nor anthropocentric, but theocentric. Humans, while in no sense owning the natural world, have a special place in it, uniquely the bearers of

God's image. This is variously expressed in strong terms of dominion or subduing, and in moderated language of working and caring for a garden.

Historically, it must be admitted, the interpretation has stressed the single aspect of dominion to excess, but the biblical texts themselves have a strong theme of restraint, with systems of checks and balances. In recent years, the relationship of humans to God's creation has been expressed most often in a recovery of Calvin's notion of the steward. God gives humans a special duty both to develop the natural world – and hence the use of technology – but also to take care of it – which puts limits on our activities. Stewardship means that humankind is answerable not merely to future human generations, but also to God, the divine owner, for how we have looked after his estate.

This is not, however, a sufficient picture, for in the biblical accounts, humans are also clearly fellow creatures in a shared creation, for example in *Psalm 104* and *Job* Chapters 38–41. Page (1986) introduces the notion of companionship to moderate the hierarchical sense of stewardship. Thus, while God puts animals under human subjugation for a wide variety of uses, these are still God's creatures first, and humans will have to give an account to God for their care of them. Old Testament injunctions such as 'Do not muzzle an ox when it is treading out the grain', 'Do not boil a kid goat in its mother's milk' (*Deuteronomy* 25:4 and 14:21), imply that wider principles of relationship set restraints on human uses.

It is not easy to put this into clear moral distinctions of use. It is much more worked out in individual expressions of overall principles. Commercial animal production by selective breeding would be allowed in general terms, but not to every degree possible. Limits are exceeded when this is taken as an end in itself, or if it becomes so dominated by other driving forces such as economics or efficiency that wider principles of God's creation are over-ridden. When taken to such degrees that harms, distortions, disablement or impairment of function begin to emerge, a good end would have been taken too far. There is a sense of balance of many other moral criteria to be held alongside the single notion of increased production.

All four categories can be critiqued as departures from a Christian base, but emphasizing or changing key elements. Ownership takes the dominion idea to extremes, following the medieval Catholic scholastic theologian Aquinas (Aquinas, 1261–64, transl. 1928) for whom rationality was the distinguishing characteristic between humans and all else in creation. It enables humans to relate to God, but animals lack this vital characteristic and exist for human benefit only. This view owed more to ancient Greek thought and the social hierarchies of the medieval world than to scripture. The worship model makes nature itself god, the creature replacing the creator. The 'what is' becomes the basis, with all the moral ambiguity that follows. The original meaning of worship as 'worth-ship' is helpful in conveying the idea of giving something its true worth, rather than, as it were, bowing down and calling it divine.

Although this fifth category has been derived from Christian ideas, it would also have a broader resonance as an intermediate position between the partnership and maintenance engineer models, for many who do not hold to a Christian faith. The maintenance engineer shares the idea of moderated use of animals, but replaces the theocentric view of the universe with an anthropocentric orientation. Looking after animals well because they are valuable to us is not sufficient, however. Animals are more than their worth to humans. Although the partnership model shares the notion that animals have intrinsic value, it overemphasizes the unity of nature, which tends to play down the validity of human intervention. Aspects which human and animal hold in common, such as both being creatures and subjects of a life or both being sentient, do not mean that humans cannot eat animals or use them for traction and carriage.

EXAMPLES OF ANIMAL INTERVENTION IN AGRICULTURE AND NOVEL USES OF FARM ANIMALS

Having identified some representative attitudes towards intervention in animals, what might these different models make of some of the recent developments in animal production? We examine a number of examples, after first considering two important contextual points.

Context 1: questioning the status quo

The first point to make is that the status quo is not an ethically acceptable or neutral ground. Far from being the recognized norm against which newer technologies should be assessed, important ethical doubts are increasingly being raised about 'conventional' animal production practices. Examples in animal breeding would include physical problems associated with rapid growth in broilers and turkeys, the hunger and frustration of feeding behaviour caused by food restriction of broiler parents and sows, and the calving difficulties of double-muscled cattle. Much too frequently, arguments are presented which assume it is sufficient simply to assert that new production technologies are not significantly different in welfare terms from what is already 'generally accepted practice'. This is not an adequate case without asking the prior question of whether the existing practice *ought* to be accepted. Comparison with a poor status quo hardly constitutes a true commendation. In a climate where current standards are being seriously challenged, each new technology should also be assessed in terms of whether it exacerbates or ameliorates the situation.

Context 2: gradualism

The need to question the status quo relates to a wider problem. One of the most characteristic features of scientific progress is a phenomenon known as gradualism. This is the effect of a series of small changes, each apparently a mere logical extension of the previous step or two, but whose cumulative effect is a radical and dramatic difference from the place one started from. A recent publication illustrates a startling difference in size between two 6-week-old chickens – the smaller bred for egg laying, the larger bred for meat production (Bruce and Bruce, 1998). They have been produced by selective breeding. If the stated aim 30 years ago was to create chickens of such size, there might well have been a public outcry. Because it happened by small stages, what was happening was not generally noticed. Only when an event brings home the effect of where our gradual progress has brought us do we stop and ask if the relentless mechanical models of both science and production which this exemplifies need some moral constraints.

Marker-assisted selection

This technique is referred to in further technical detail in Chapter 2. Samples of animal tissue or blood are analysed for particular genetic markers, which are associated with some trait of interest to a breeder. The results of the analysis are used to decide which individual animals are allowed to reproduce the next generation. It is primarily a tool to assist selective breeding, and as such raises the same ethical issues about both existing and future practices, but it can accentuate them. In so far as marker-assisted selection can bring about genetic changes more quickly, then both potential negative as well as positive effects may also be achieved more quickly.

This development is chosen because it is an illustration of something which would only be regarded as intrinsically inappropriate use of an animal under the 'worship' model. Under this notion, it could be seen as wrong in principle to ask the animal to provide us with anything, and so selective breeding itself would be problematical. If animals represent an element of the divine, it is inappropriate to use a production metaphor at all. In this case, marker-assisted selection is unlikely to be acceptable, because it would be manipulating the reproductive processes of the animal simply as a means to an end. Would the end of alleviating disease in the animal be allowed, or is even this tampering with the 'givenness' of things?

If in the partnership model we might use an animal for purposes which impinge only very slightly on the animal itself, selective breeding and the use of markers to assist it might be seen as acceptable interventions in principle. In many of the production contexts in which it would be used, however, serious objections would probably be raised about the ends for which it was being done, or about the degree of intervention involved. The

Christian model would ask whether a sense of balance had been lost, in so far as the status quo in animal production indicates some areas of significant abuse. It would suggest that the production motive had lost track of wider obligations to animals as God's creatures. Marker-assisted selection would thus not be appropriate where it might spread, exacerbate or accelerate existing problems. The ownership model would disregard such constraints as irrelevant, and the maintenance engineer would only see problems if the technique was used to produce animals with serious welfare problems.

Genetic engineering of animals in principle

The genetic engineering of animals has stimulated much public discussion, and raises a number of important questions about human intervention in animals. Indications to date are that this is less relevant to conventional animal production than to novel uses. Genetic engineering in animals is not so much about cheap food and lower production costs, as diversification into new fields. We will discuss three examples: pharmaceutical production in the milk of farm animals, xenotransplantation using pigs, and the possible use of farm animals as models of human disease. First, however, we will consider the underlying ethical question of whether it is right to modify animals genetically at all, and aspects of animal welfare to do with the process of modification.

Is genetic engineering fundamentally wrong, irrespective of its application? This is an example of an intrinsic ethical issue, something right or wrong in itself, regardless of its consequences. Although the validity of such questions is sometimes dismissed in scientific circles as emotional or irrational, especially in matters of public policy, the Banner Report on the ethics of emerging technologies in animal breeding stressed the importance of taking intrinsic objections into account in relation to animal technologies (Banner, 1995). It has also been pointed out that such views are no less 'rational' than those constructed upon the assumptions of scientific rationality (Bruce and Bruce, 1998).

One of the frequent accusations about genetic engineering is that it is 'playing God'. From a secular perspective, this is a shorthand for a general sense that this is doing something that is not for humans to attempt – either something forbidden by God or some less defined sense of absolutes, or else too risky. For Christian theology, this secular use of a theological term raises a smile, not least because there is also a positive sense of playing God in the creative and caring way humans are encouraged to act in nature on God's behalf or, as some scientists of Christian faith earlier put it, 'thinking God's thoughts after him'. The question is whether manipulating genes and transferring them amongst widely varying species is a proper expression of human creativity in science, or is infringing some fundamental limits in the nature of things.

In our categorization, the worship model would indeed take the latter position. If nature is divine, and animals are a reflection of divinity, then to alter their basic building blocks by genetic manipulation would be a fundamentally unacceptable interference. Some religious believers in various traditions might consider that to change a single gene in an animal would be attempting to change God's best design. A partnership model might argue that we should not genetically engineer animals in any way that humans would not do to each other. This would not mean a wholesale rejection of the technique, but serious questions would be asked about the specific uses. The 'No, unless' approach might allow use where the prime benefit was to the animal, such as increased disease resistance, or in cases where a major human benefit could be achieved with minimal interference in the animal, but probably not for increased growth rate in animal production. The idea of swapping genes between species would be regarded with great scepticism, upsetting the wisdom inherent in the natural order by humans who did not know the full extent of the unprecedented changes they were making.

Neither the maintenance engineer nor ownership models would find problems in principle with genetic engineering of animals. Both would require enough data to be sure no harm would result to humans, and, for the maintenance engineer, serious harm to the animal also. The middle or Christian viewpoint would be akin to the maintenance engineer, but would be more precautionary about the sufficiency of human knowledge and wisdom in a relatively new field of research. It would be more critical about what is a permissible level and nature of intervention, what motives were driving it, and it would ask if there were better ways to the same end without manipulating the animal.

Special issues relating to transgenesis

Transgenesis – mixing genes between widely differing species – opens up some special questions about the significance of DNA in relation to the nature of an animal. This depends on how far the delineation of species and the exact genetic make up of a particular creature were regarded as God-given or evolved norms from which humans are not to depart. Some religious interpretations would consider these to be God-given, and so render transgenesis forbidden. Others, for instance the version of the Christian model set out above, would argue that the nature of an animal is more than just the sum of its genes but rather is the essence of the creature as a whole. To change one or two genes would not, therefore, constitute a change to some blueprint which would irretrievably violate the animal, unless, consequentially, it brought about a severe impairment or suffering to it.

Particular types of gene may have special significance. Although interpretations may vary, in some religious traditions the presence in an organism of a gene from a ritually unclean animal, such as a pig in Islamic and Jewish contexts, could render the organism unclean. Similarly, vegetarians might find the use of any genes of animal origin unacceptable in food. Some meat eaters express concerns that eating meat from an animal with a human gene might constitute a sort of cannibalism, or at least represents breaking an inherent barrier which should not be breached.

In the UK, the Polkinghorne Committee (Polkinghorne, 1993) pointed out that any gene transferred to another organism will have been copied millions of times, for example via bacteria (Chapter 2). They argued that it is no longer a human or animal gene, but a copy. The validity of this argument has been challenged, however, as too reductionist. A human gene remains human by association, regardless of being copied. Moreover, the reason that a human gene is being used is because its characteristic human function is desired in another organism. It is thus the information coded by a gene that is important, and it is therefore immaterial whether a particular DNA molecule literally came out of a human, because the information is from a human (Reiss and Straughan, 1996; Bruce and Bruce, 1998). Attitudes to this question are likely to depend on whether the key criterion is the physical origin of the DNA or the 'meaning' coded in it.

Genetic engineering and animal welfare

The partnership, Christian and, to a lesser extent, maintenance engineer models would be concerned about the welfare of the animal as a result of genetic engineering. Appleby has identified several stages at which this needs to be considered (Appleby, 1998).

The effect of genetic modification techniques themselves

The techniques used in genetic modification can cause welfare problems irrespective of the nature of the modification achieved, and in some cases these may be the more important issue. Handling animals during the procedures may frighten them. They may need to be isolated and may be kept in barren surroundings. At this level, the issues are no different from those related to normal husbandry practices, but there may be other consequences more specific to genetic manipulation techniques. Hormone injections are used to achieve superovulation, and surgery is often necessary for embryo retrieval and transfer. All of these can result in welfare problems for the animal. The partnership model would be reluctant to regard harmful interventions of this nature as justifiable, and this might preclude an otherwise acceptable modification. The Christian model could tolerate a level of harm, in proportion to the benefit. The maintenance engineer would look more to the end intended, but would also stop if the technique involved an undue level of harm.

Direct, side and unintended effects of genetic modification

The result of a genetic change need not be adverse welfare. It may be neutral or even positive, as in the case of seeking to increase an animal's resistance to disease. Intentional negative effects are common, however, in animals used to study human diseases or to study developmental processes, for example mice genetically engineered to have cystic fibrosis. Such negative effects are often severe because the diseases being studied are very serious. This is indeed the justification for causing the negative effects. In addition to these intended effects, there may be unintended side effects of genetic procedures. The commonly quoted but now rather dated example is the Beltsville pig, also referred to in Chapters 2 and 5, which was modified to include the human growth hormone gene but which suffered very severe joint problems. Much the same judgements would be made in the different categories as above, but a wider question is asked about limits to harm.

Perspectives on acceptable and unacceptable harm

In the UK, the Banner Committee review of the ethics of emerging technologies in animal breeding delineated three basic principles, which could be seen as broadly consonant with what we have defined as the Christian or intermediate model (Banner, 1995). There are some harms to animals that should never be done, whatever the purpose, a recognition of intrinsic worth. Other harms may be allowed, but must be justified as being outweighed by the benefit which realistically may be expected, and should in any case be minimized. The difficulty is in how to decide what kinds or degrees of harm we must not inflict. The Banner Report cites an imaginary situation of genetically modifying pigs by reducing their sentience and responsiveness in order to improve efficiency of food conversion. It argues that this would be unacceptable in principle because it would result in treating an animal as a means to an end only, regardless of the ends and purposes natural to the animal. On the other hand, another imaginary genetic modification, this time in order to produce poultry which only produce female chicks, is not seen as morally significant because this modification would not deprive the chickens of the freedom to express normal behaviour. Perhaps not everyone would agree with the distinction, but this does give some guidelines on how such decisions could be taken.

The recent UK Farm Animal Welfare Council (FAWC) report on the welfare implications of animal cloning (FAWC, 1998) takes these principles and develops them. It agrees with the Banner Committee that a procedure may be considered intrinsically objectionable if it generates animals with severely reduced sentience. The FAWC report also suggests that a procedure which inflicts very severe or lasting pain on the animals concerned should fall into this category. It rejects in the same manner any procedures which involve an unacceptable violation of the integrity of the living being, or involve the mixing of kinds of animals to an extent which is not acceptable. The report does not define either of these latter factors, unfortunately. Examples which might be suggested include animals such as the sheep–goat chimera known as a 'geep', and the introduction of the nucleus of one species into the cytoplasm of another.

Having considered whether animal genetic engineering could be acceptable in principle, and some generic issues of animal welfare, we now turn to three examples, which illustrate the different types of conclusions which arise from the five categories, with the exception of the worship model which would find genetic engineering unacceptable in principle, and so is not considered further in the following sections.

Genetic engineering to produce pharmaceuticals in sheep's milk

The ability to modify animals genetically in order to produce valuable proteins such as pharmaceuticals in their milk has been one of the most innovative applications of genetic engineering techniques. It has been applied to cattle, sheep, goats and rabbits in order to produce a variety of different proteins. The example considered here is the one furthest ahead in development, that of α_1-antitrypsin for treating lung diseases, which is produced in the milk of sheep by adding the human gene which codes for the protein in humans. High-value pharmaceuticals could offer attractive alternatives for animal producers to meat, wool and milk production, especially in parts of the world with very good animal health status.

Of all animal applications of genetic engineering, this seems to be one of the least controversial. The use of sheep's milk is traditional and, therefore, to produce a particular protein in the milk would not seem an undue departure from the current situation, particularly since the sheep version of α_1-antitrypsin is produced by the animal in any case, albeit in the liver rather than in milk. The Christian and maintenance engineer models would find this acceptable. In the partnership model, the intervention in the animal is judged to be small, the human medical need being addressed is considerable, and other routes to obtain the protein are much more difficult. Indeed, it could be argued as a genuine partnership, in which humans give especial husbandry and care of the sheep in exchange for a valuable product in the sheep's milk.

Investigations to date have shown no welfare concerns from this particular example (Appleby, 1998). Other research to produce a more active protein erythropoietin showed unacceptable welfare effects for the animals, which led to the trials being terminated. All but the ownership model would recognize that there are limits even to a relatively benign type of intervention.

Genetically modifying pigs to produce organs for transplant into humans

A much more serious intervention in farm animals is xenotransplantation. Various pig organs have potential for transplantation into human beings. Pigs are being modified genetically to try to overcome rejection by the human immune system. The ownership model and maintenance engineer would consider that the huge possible human benefits make the technology desirable. The technology is driven by hope that this would meet the short-fall in supply for human organs, where people currently are dying while on the waiting list. The technology is seen as just an extension of the existing use of pig heart valves, since human utility is the prime consideration. The 'ham sandwich' argument is cited: if we accept killing a pig for food, it is surely right to kill it to treat a fatal human disease. Both models could draw back on the grounds of high safety risks to humans, but are unlikely to do so over its effect on animals.

The partnership model disagrees with this utilitarian, anthropocentric view. In an ecocentric view of the world, where other species have a very high claim for our respect, how can this be an acceptable relationship with our fellow creatures to breed and engineer them for spare parts? Unlike eating animals, there is no parallel to xenotransplantation in nature. More-over, using a live animal heart is a very large step from using an inert heart valve tissue for its elastic properties. The fact that xenotransplantation is unnatural may not necessarily make it wrong, but it prompts the question of whether this is an acceptable extension of human use of animals from traditional suppliers of food, clothing, traction, transport and manure. Quite apart from unknown risks involved in such a major change, this model would conclude that this is a wrong way of using an animal.

The Christian or middle model would be torn both ways, with a deep sense of concern for the human suffering that might be alleviated, but a deep reluctance to treat another of God's creatures merely as a source of spare parts. The pressures arising from the success of transplant procedures are not the only criteria. Unlike human heart transplants, where the donor is already a cadaver, this means serious intervention in a live animal which will destroy its life. It also raises questions of how acceptable would be the quality of life for the pigs kept, of necessity, in a highly sterile environment? To justify this intervention requires there to be a high expectation of success with a prolonged, much enhanced quality of life for the patient. It is justified to conduct research while this remains a realistic prospect. However, given the complexity of the multiple genetic modifications that are now likely to be needed, it is not a foregone conclusion that there will come a point where that ethical balance would be reached for it to become an accepted therapy. This viewpoint would also ask what other avenues might be pursued to achieve the same end, and at what point there eventually comes a limit to forever replacing human organs as though it would make us immortal.

Sheep as a model for cystic fibrosis

The largest application of transgenic animals by far is the controversial use of mice as models of human disease and for tests for potential therapies. Hitherto this has not impacted much on farm animals. Studying a larger animal could be much easier, however. A proposal from the Roslin Institute to develop a sheep genetically modified to exhibit the human symptoms of cystic fibrosis focuses the wider question of how far, if at all, humans are entitled to intervene in animals knowingly to cause them suffering, and in which types of animals. It represents the extreme end of a progression of degrees of intervention in animals to benefit humans.

For the ownership model, there is no issue to debate. A clear and deeply serious human health problem could be alleviated by using animals. The worship and partnership models would argue that deliberately to give a fellow creature a fatal human disease departs from any sense of acceptable human behaviour. This would be one of Banner's 'harms that should on no account be allowed'. For the Christian model, the dilemma of xenotransplantation is taken one stage further, and even the maintenance engineer might now also have qualms. Were it not for the seriousness of diseases such as cystic fibrosis and cancer, one would not contemplate using even mice in this way, but what about sheep? How do we decide amongst animal species?

The Nuffield Council on Bioethics (1996) gave good reasons to reject using primates for xenotransplantation, but accepted using pigs without establishing what made the crucial difference, given for example pigs' intelligence and physiological analogies with humans. At the other end of the mammalian spectrum, the fear has been expressed that because the medical research culture permits mice to be used, mice become mere catalogue items, and no longer animals (Bruce and Bruce, 1998). In the UK, the 'three R's' principle is used as a guideline – to reduce, refine, and as far as possible replace the use of animals for experimentation. In the light of this, it would be necessary to ask how good is the model and how necessary is the intervention, in order to satisfy the 'No, unless' policy. There is a case to address that farm animals are the critical point where utility to human medicine might have reached a limit.

Animal cloning

The overwhelming and somewhat self-perpetuating media fascination with Dolly and speculations about human cloning have overshadowed the significant ethical issues which are already presented by animal cloning. The following insights stem from several years of interaction with the Roslin Institute researchers over ethical aspects of cloning, and attempts to redress the balance of focus (Bruce, 1997, 1998). The FAWC report mentioned above was a welcome input to a neglected issue (FAWC, 1998). Cloning occurs naturally in many plants and microorganisms, and in some lower animals. However, it does not normally happen in mammals, except for the occurrence of 'identical' twins. There is first of all a fundamental question. Should we respect this biological distinction or celebrate our capacity to override it? Secondly, under what circumstances might or might not animal cloning be done, according to the different criteria?

Cloning animals in principle

To the first question, the worship and ownership models represent, respect-ively, the two poles of the answer, but what of the shades in between? Two of the principal objections to cloning human beings are the control given of one human over another's genetic make-up, and the instrumental nature of most of the reasons postulated for doing it. In themselves neither of these would constitute an overwhelming reason against cloning animals, if a certain amount of valid human use of animals is granted, except under the worship and possibly partnership models. Both would object in any case to the invasive procedures necessary to achieve animal cloning, whether by nuclear transfer or embryo splitting.

For creatures that rely on sexual reproduction it is important for a healthy population to maintain good genetic diversity. To force asexual reproduction on to such animals could be said to be a step in the wrong direction, going against the grain of evolutionary variety in nature. The maintenance engineer might therefore be constrained to avoid any adverse effects which could impact on production. There is an important religious dimension to this point. The very diversity of nature is seen to be fundamental to the divine purpose, and, for example in Jewish, Islamic and Christian scriptures, a cause of praise to its creator. Diversity in nature is also a source of pleasure and usefulness to humans. Where God evolves a system of boundless possibilities by diversification, should humans think twice before selecting out certain functions we think are the best, and simply replicating them wherever we see fit to do so? This is not so much an absolute limit, but a check in the process. Rather than go to the opposite extreme of automatically adding cloning to the set of technological manipulations we already carry out on farm animals, under the ownership or maintenance engineer logic, the intermediate model would want first to ask whether such an intervention is justified by the reasons for doing it. The Dutch transgenic animal committee and the European Commission's Group of Advisors on Ethics in Biotechnology recommended just such a 'No, unless' approach to this question (European Commission, 1997). We now look at several examples.

Cloning to improve genetic modification in farm animals

The main aim of Roslin's research programme was not cloning for itself, but to find better ways of genetic modification for pharmaceutical production in farm animals' milk. Instead of the very hit and miss process of micro-injection, they wished to use fewer animals if a method could be found to grow an animal from selected genetically modified cells. The work produced first Megan and Morag as cloned sheep from embryonic tissue, then the genetically modified cloned sheep, Polly. That she was also a clone was something of a side effect, and natural breeding would be used thereafter. In an early evaluation of this application, the Church of Scotland accepted animal cloning in this limited context, aimed at a clear medical need, reducing the number of animals used, where cloning was not the main intention, and where natural methods would not work (Church of Scotland, 1997).

Following this original example, the method is capable of many wider applications (Wilmut *et al.*, 1997). It could be used in other areas of animal genetic engineering such as xenotransplantation, but it also provides a way to perform new types of genetic modification not before possible in farm animals (Wilmut *et al.*, 1997). Alternatively, a successfully genetically engineered animal might be cloned to provide more founder animals, as in a recent Dutch proposal to clone transgenic cows but not bulls (*New Scientist*, 1999). The surprise discovery that mice can also be cloned opens up many further possibilities for research and thus for developing other applications in animals. While Roslin's original work in sheep cloning for pharmaceutical production would be acceptable under the intermediate model, other transgenic applications of cloning would need examining on their own merits, and some might present more problems. The application to humans is quite another story. Criticisms that Roslin should never have started down this road, given the risk that it might end in human cloning, are easier to make in retrospect than when faced with a speculative research proposal to do something considered biologically impossible.

Welfare questions

There is also a generic problem of animal welfare considerations in cloning. Much of the basic science of nuclear transfer remains to be understood. With such novel technology, the FAWC (1998) was cautious over the potential uses of cloning in animals. It asked for regulations to protect cloned farmed livestock and a National Standing Committee to oversee the development of cloning technology. It called for a moratorium on nuclear transfer cloning in commercial agriculture while further investigation is made of welfare problems and uncertainties over oversized offspring, perinatal and birth problems, and aged DNA. Both the maintenance engineer and Christian positions would have valid concerns at this point. There is, however, a problem in how far such welfare investigations should be pursued if they may themselves cause harm to the animals involved.

Cloning in farm animal production

Cloning technology could also be applied to conventional farm animal production for meat and milk, not so much in sheep as in cattle and pigs. Cattle have already been cloned both by nuclear transfer cloning and the older method of embryo splitting. Most dairy cattle in the UK are already produced by artificial insemination, where the semen from one select bull can service numerous cows, and embryo transfer extends this further. It might be the next logical step to clone prime cattle in a breeding programme, to raise more breeding stock to the highest level of 'genetic merit', or even to clone the best beasts for fattening for slaughter. For the maintenance engineer model, this would cause no difficulties, but for the Christian or middle position, this could raise problems. The Church of Scotland General Assembly made a distinction between this case and pharmaceutical production (Church of Scotland, 1997). Against a context where existing animal-breeding practices have raised serious questions about the dominance of the production motive, it took the view that to clone animals routinely for meat or milk production would be taking instrumental intervention into animals one step too far, given that natural methods of breeding exist. There are limits on how far humans should commodify animals for their functional worth. Copying the complete genetic blueprint for efficiency's sake is a factory mass production mentality inappropriate to animal husbandry. Our fellow creatures are more than identical widgets on an assembly line.

Other applications of animal cloning have been speculated upon, including combating animal disease and even the protection of endangered species from extinction. Unlike the previous example where the primary benefit appears to be production efficiency and convenience, rather than human or animal need, animals would be the beneficiaries. Here the Christian and partnership models would probably have less objection.

Societal aspects of animal use

In addition to the ethical and philosophical aspects so far considered, there are also important social questions to be examined. A common thread throughout is the role of science and economics as driving forces of the developments. To what extent is this leading to challenges or improvements to animal welfare?

We have already commented on cases where an overemphasis on the imperative of production, driven by economic considerations, has lost the wider considerations with regard to animal use and welfare. The pressure to reduce overheads to remain competitive draws demands for ever greater efficiency. However, this concept of efficiency is taken from the sphere of mechanical engineering, and beyond a certain point this starts to sit very awkwardly with systems based on living organisms. The biosphere furnishes excellent modes of efficiency, but in a very different sense. Perhaps efficiency models should be drawn more from these rather than from the factory.

The scientific culture can itself lead to something of a tension. To the extent that animal production is being influenced by animal science, it should be aware of the culture generated by the scientific mindset. Two key concepts of pure science are its stress on the disinterested observer, and its method of reducing complex phenomena into simple functions, which can be examined, understood and then manipulated at will. Together, these create a powerful but clinical way of investigation. The further it focuses on manipulating individual parts, the harder it may be to relate the parts back to the whole. When applied to the context of animals, one must be aware of the danger of pushing wider criteria to one side. Left to its own logic, science will tend to push towards an ownership model, especially when research is driven increasingly by commercial values and priorities. It is therefore up to us to work at how to harness this power without losing sight of the context, both in relation to the animals themselves and also our own humanity.

Lastly, there is the issue of public accountability of developments in animal production. Public concerns in this matter vary greatly among different countries and parts of the world. In the West, there is a considerable social separation of animal production from the awareness and understanding of the public. Practices which have long been commonplace in a production context, such as castration and embryo transfer, may be quite shocking to the general public. Novel methods appear in the media, with the public having had little say in their acceptability. Animal pressure groups may take 'direct action' that is representative of only a small minority. There is an urgent need to address this democratic deficit. The further methods such as genetic engineering and cloning are taken, the more they are likely to challenge social values of what it is acceptable to do to animals. The appropriate response is surely to ensure a correspondingly greater involvement of the public in decisions as to what is being done, supposedly, in its name.

Conclusions

We have considered a variety of applications of biotechnology to farm animal use and production, against a spectrum of attitudes and relationships to nature. These models are to some extent artificial, like peaks in a statistical population which in practice merge into each other. They do, however, serve to show some of the different ways in which it is possible to approach animal use and welfare. They can help us to clarify why we think what we think, and what underlying motives and values drive our corresponding actions. We have our own preferred model. Recognizing where we each come from is important in understanding ourselves intellectually, morally and spiritually, and, if necessary, changing our actions.

In Europe, we are seeing, on the one hand, increasingly technological ways of seeing animals, and, on the other, a general belief that, if anything, we need to be reducing the amount of animal use. If our relationship with nature is already wrong because we are exploiting too many animals, how do we handle a new technology which exploits animals further? The Nuffield Council on Bioethics' (1996) comment about xenotransplantation has a wider applicability:

> How far should public policy be based on concerns about the underlying attitudes that the development of a particular technology ought to reveal? If human beings have not got their relationship with the rest of nature right, and there is continual unjustifiable exploitation of other species for human use, then it would be wrong to allow the development of a new technology that increased this exploitation ... what sort of people do our social and technical practices reveal us to be? If we do not like what we see when we look honestly in the mirror, then there is cause for thought at least. (para 4.22)

Many factors can be summed up in the way we use language regarding the new developments in animal genetics. Terms such as 'bioreactor' for production of pharmaceuticals in milk, 'spare part supplier' and 'genetic engineering' itself are all borrowed from the sphere of industrial production of inanimate objects, not living animals. If we now accept describing an animal primarily by its functionality, as a means to an end, our mental conception of the animal can become reduced to the merely instrumental. The maintenance engineer and owner may well be comfortable with this tendency, but the Christian, partnership and worship models would profoundly object. A change in terminology does not itself ensure that the animal is badly treated, but, by extrapolating from concepts of industry, a statement is made that the animals are more closely related to the non-living world than the living.

Given that genetic technology is itself an intrinsically reductionist approach, if our relationship to animals says something about our humanity then we need to work hard to ensure that we relate to the animal as an animal not as a machine.

REFERENCES

New Scientist (1999) Cut the bull. *New Scientist,* 4 February, 2172, 5.

Appleby, M.C. (1998) Genetic engineering, welfare and accountability. *Journal of Applied Animal Welfare Science* 1, 255–273.

Aquinas, T. (1261–64) *Summa Contra Gentiles.* Third Book, Part II, Ch.CXII (English translation by the Dominican Fathers, Benzger Brothers 1928, New York).

Banner, M. (Chairman) (1995) *Report of the Committee to Consider the Ethical Implications of Emerging Technologies in the Breeding of Farm Animals.* Ministry of Agriculture, Fisheries and Food, HMSO, London.

Bruce, D.M. (1993) *What Are We, by Nature?* Presented at the Centre for Philosophy, Technology and Society Conference on Philosophy and Environment, Aberdeen University, 13 September 1993.

Bruce, D.M. (1997) A view from Edinburgh. In: Cole-Turner, R. (ed.), *Human Cloning: Religious Responses.* Westminster John Knox Press, Louisville, Kentucky, pp. 1–11.

Bruce, D.M. (1998) Polly, Dolly, Megan and Morag: a view from Edinburgh on cloning and genetic engineering. *Society for Philosophy and Technology Journal,* 3, http://scholar.lib.vt.edu/ejournals/SPT/v3_n2html/BRUCE.html

Bruce, D.M. and Bruce, A. (eds) (1998) *Engineering Genesis.* Earthscan Publications, London.

Church of Scotland (1997) Board of National Mission Supplementary Report on the Cloning of Animals and Humans. In: *Supplementary Reports to the General Assembly and Deliverances of the General Assembly 1997,* page 36/22, and Board of National Mission Deliverance number 35, p. 16 of the Deliverances.

Duncan, I.J.H. and Fraser, D. (1997) Understanding animal welfare. In: Appleby, M.C and Hughes, B.O. (eds), *Animal Welfare.* CAB International, Wallingford, UK, pp. 19–31.

European Commission (EC) (1997) Group of Advisors on the Ethical Implications of Biotechnology, *Opinion on Cloning of Humans and Animals by Nuclear Transfer.* European Commission, Brussels.

FAWC (1998) *Report on the Implications of Cloning for the Welfare of Farmed Livestock.* Ministry of Agriculture, Fisheries and Food, London.

Nuffield Council on Bioethics (1996) *Animal-to-Human Transplants: The Ethics of Xenotransplantation.* Nuffield Council on Bioethics, London.

Page, R. (1986) The Earth is the Lord's: responsible land use in a religious perspective. In: *While the Earth Endures.* SRT Project, Edinburgh, pp. 1–15.

Polkinghorne, J.C. (Chairman) (1993) *Report of the Committee on the Ethics of Genetic Modification and Food Use.* Ministry of Agriculture, Fisheries and Food, HMSO, London.

Regan, T. (1988) *The Case for Animal Rights.* Routledge, London.

Reiss, M.J. and Straughan, R. (1996) *Improving Nature.* Cambridge University Press, Cambridge.

Singer, P. (1990) *Animal Liberation.* 2nd edn. Cape, London.

Thatcher, M.H. (1990) Quoted in *This Common Inheritance: Britain's Environmental Strategy,* para 1.14, HMSO, London.

Visser, M. and Verhoog, H. (1999) *The Moral Relevance of Naturalness in Discussions on Genetic Manipulation of Animals.* Netherlands National Science Organization, Amsterdam.

Wilmut, I., Schnieke, A.E., McWhir, J., Kind, A.J. and Campbell, K.H.S. (1997) Viable offspring derived from fetal and adult mammalian cells. *Nature* 385, 810–813.

Wilson, P. (1998) Case study on bovine somatotropin. In: Bruce, D.M. and Bruce, A. (eds), *Engineering Genesis.* Earthscan Publications, London, pp. 55–59.

Agribusiness and Consumer Ethical Concerns over Animal Use and Foods of Animal Origin: the Emergence of New Ethical Thinking in Society

4

Bernard E. Rollin

Director of Bioethical Planning, Colorado State University, Fort Collins, Colorado, USA

ETHICS AND SOCIETY

It is historically indubitable that social ethics determine to a significant extent consumer acceptance and rejection of goods and services provided by business as well as how business comports itself. Social moral concern about sweat shops, child labour, mistreatment of workers in the developing world, dangerous working conditions, union-busting businesses, exclusion of women and minorities from professions or well-paying positions, corporate environmental policies and animal testing, among other concerns, have forced business to change their policies in these and myriad other areas. As the recent cases of McDonalds, Nike, the Gap and the Body Shop dramatically illustrate, the connection between social ethics and consumer acceptance of products is not merely of historical interest, but is very much alive.

Indeed, it is difficult to see how anyone with even a modicum of understanding of the role of ethics in society could miss this point. Even in antiquity, Plato stressed that our vision of the good informs all other human activities and decisions individually and socially. In other words, our views of what is right and wrong, good and bad, just and unjust, fair and unfair constrain and direct what we consider acceptable and unacceptable in all aspects of personal and collective life.

Business and science

However, this manifest point *has* been overlooked consistently by the two great institutions, business and science, that have shaped modern life since the Renaissance (Rollin, 1998). One can speculate why this is the case. In the first place, both of these institutions embodied rejections of the medieval feudal–theological synthesis that grounded both knowledge and behaviour in fixed, religiously based institutions. With the rise of science and capitalism (and democracy and Protestantism), the tie to ultimate authority and fixed institutions was broken. Each person could (and was obliged to) seek the truth on his own, pursue his own interest and vote in his own interest. The inevitable scepticism about established authority in knowledge or ethics devolved into sceptical positivism, which implicitly (and later explicitly) questioned the traditional basis for absolute ethical judgements. In science, this led to the dictum that science was 'value-free', a notion explicitly trumpeted in the 20th century. In business, similar scepticism reigned, resulting in a tendency to accept only efficiency, productivity and profit as legitimate business values, with ethics seen as 'opinion', 'subjective' and non-quantifiable.

Contemporary values in western society

This fundamental tendency to scepticism has been buttressed by recent developments in political life, notably the rampant emphasis on the rights of individuals and subgroups of society, with universal communitarian values submerged. If anything, ethics came to be seen as itself a free market of competing values, with what 'sells' determining what is ethical, e.g. yesterday homosexual behaviour was criminal, whereas today it is to be respected as expressing an alternative lifestyle. In this way, radical ethical pluralism, always implicit in democracy, can further perpetuate the belief that there is no rational way to talk about ethics.

Thus business has tended to ignore ethical thoughts and reflection until it becomes impossible to do so, that is until society hits business over the head with ethical decisions, such as the unacceptability of child labour or hazardous working conditions or sexual harassment. Business rarely *anticipates* ethical change, and almost always reacts to it. These reactions are often knee jerk and non-reflective, as shown by the lengths to which some business corporations go in hiring token minority group members; or the witch hunts that can turn innocent remarks into actionable cases of sexual harassment. In short, business tends to adopt the current fashion in ethics, without any real understanding of the underlying rational bases for social ethical change, and with style taking the place of substance.

The failure to understand the rational basis for ethical change and the dynamics of that change can lead to real losses for commerce and industry. If business does not see social ethical change coming, for example, it may make expensive investments incompatible with that change, which must later be substantially revised, costing significant amounts of money. For example, the confinement swine industry in the USA has been oblivious to growing social ethical concerns about environmental despoliation and has built units, for example in Colorado, that must be substantially modified in a very costly way to meet the results of new legislation designed to protect the environment.

Recent ethical issues facing business and science

There are various other ways in which business can lose when it ignores social ethical change or concerns. Nike's labour practices in China provide an excellent warning to other industries. An even more eye-opening example may be found in the rejection of biotechnology in Europe. It is well known that Europe is much more negative towards biotechnology than is the USA. For example, bovine somatotropin (bST) has not been allowed in Europe; neither has the patenting of animals. A variety of theories have circulated regarding that rejection, most of them focusing on the claim that Europeans are more risk-aversive than are Americans, given the frontier ethic of the latter, the fact that there has never been a foreign war fought on US soil, etc. The standard wisdom as articulated by regulators and business people is that Europeans reject biotechnology because they are more concerned about possible risks associated with that technology. However, recent research calls that unchallenged assumption into question.

European views on issues in biotechnology

In June 1997, a team of researchers working as part of a Concerted Action of the European Commission, and coordinated by George Gaskell of the London School of Economics, released the results of a survey of public attitudes towards biotechnology conducted in each of 16 European Union countries. According to Gaskell (personal communication), the results astonished the researchers, shattering both their preconceptions and conventional scientific wisdom about social responses towards biotechnology. The researchers found that 'few [people] approve of the use of transgenic animals for research' (Gaskell, 1997). In addition, 'there is a striking mismatch between the traditional concern of regulators with issues of risk and safety, and that of the public, which centres on questions of moral acceptability' (Gaskell, 1997).

Although conventional wisdom suggests that the overwhelming social concern about biotechnology is risk, the survey confuted that presupposition. When the 17,000 people surveyed were asked about six different aspects of biotechnology – genetic testing (using genetic tests to detect heritable diseases); medicine production, using human genes in bacteria to produce medicines or vaccines, as has been done with insulin; crop plant modification, e.g. moving genes from plant species into crops to produce resistance to insects; food production, e.g. to make foods higher in protein or have longer storage life; transgenic research animals genetically modified for research, such as the oncomouse; and xenotransplants, introducing human genes into animals to render their organs immunocompatible for human transplants – all were perceived as potentially useful, but the uses of transgenic animals for research and transplantation were seen as *morally unacceptable*. Gaskell (1997) reports as follows:

> The pattern of results across the six applications suggests that perceptions of usefulness, riskiness, and moral acceptability could be combined to shape overall support [for biotechnology] in the following way. First, usefulness is a precondition of support; second, people seem prepared to accept some risk as long as there is a perception of usefulness and no moral concern; but third, and crucially, moral doubts act as a veto irrespective of people's views on use and risk. The finding that risk is less significant than moral acceptability in shaping public perceptions of biotechnology holds true in each EU country and across all six specific applications.... This has important implications for policy making. In general, policy debate about biotechnology has been couched in terms of potential risks to the environment and/or human health. If, however, people are more swayed by moral considerations, public concern is unlikely to be alleviated by technically based reassurances and/or regulatory initiatives that deal exclusively with the avoidance of harm.

Regrettably, the study does not enumerate or address the specific moral concerns which rendered the creation of transgenic animals for research morally unacceptable. In what follows, I shall briefly attempt to provide a plausible rational reconstruction of justifiable social moral concern about the production of transgenic animals for biomedical research, and thereby illustrate the power of social ethics in determining what sorts of behaviour, even socially and economically advantageous behaviour, are allowed to proceed.

Primary and secondary senses of ethics

Before embarking upon this specific task, it is necessary to review some basic notions about ethics. It is essential, first of all, to distinguish between ethics$_1$, the primary sense of ethics, and ethics$_2$, a secondary sense. Ethics$_1$ is the set of beliefs about right and wrong, good and bad, fairness and unfairness, justice and injustice that individuals, societies and subgroups of society would agree guide their actions. Ethics$_1$, is imparted by many teachers – parents, church, schools, peers, media, etc. Ethics$_2$ is the rational examination of ethics$_1$ – the attempt to validate ethics$_1$ claims, the drawing out of unnoticed implications from ethics$_1$ principles, the attempt to render consistent putatively inconsistent ethics, principles such as 'don't take any guff from anyone' and 'turn the other cheek'. As such, ethics$_2$ is a philosophical or conceptual activity, not a form of preaching. Indeed, the subject we are engaging in this paper is an ethics$_2$ activity.

Personal and social ethics

One fundamental ethics$_2$ point that must be stressed at the outset in the face of the positivistic scepticism alluded to earlier is that ethics$_1$ is not simply a matter of personal opinion. Fundamental judgements about the moral acceptability of behaviour that has a major and direct impact on others are not left to individual opinion in society. If anyone doubts this, let them flagrantly rob a bank and, when they are caught, argue that, in their ethical opinion, bank robbery is morally acceptable if one needs the money! Clearly, issues with implications for the well-being of others are dealt with by the *consensus social ethic*, which one finds fairly well articulated in the legal system (though not perfectly). All laws and regulations, from the siting of pornographic book stores away from school zones to laws against murder, rape, insider trading or sexual harassment, are based in moral insights. Without a social consensus ethic, social interaction would be impossible as chaos and anarchy would rule.

Not all activities with ethical import, however, are regulated by the social consensus ethic as some are left to one's *personal ethic*. Such questions as what religion one belongs to, what one eats, what if any charity one gives are left to one's personal ethic in today's western societies, though not of course in all societies historically.

Two points need to be stressed at the outset. First of all, when we point out the need for a social consensus ethic, we are not *ipso facto* committed to *ethical relativism*, the belief that all social ethics are equally correct or desirable. In fact, I have argued vigorously against this claim, utilizing the strategy that, given the job of a social ethic, we can say that the democratic ethic does the job better than a totalitarian ethic that allows arbitrary seizure of life and property. However, that is not germane to our discussion here.

Second, neither the social ethic nor the personal ethic is fixed and immutable. Things move in and out of the personal and social ethics. For example, prior to the 1960s, selling and renting of property, hiring and firing of employees were left to one's personal ethic in the USA, and were seen as paradigmatic examples of personal choice. Today, however, such activities are strictly constrained by the social ethic. Generally, things move from the personal ethic to the social ethic when leaving them to the personal ethic is perceived as generating widespread injustice or unfairness. For example, leaving the selling and renting of property to people's personal ethic resulted in great injustice to minorities. Thus, if individuals or subgroups or society wish to preserve their autonomy, they must assure society that their behaviour is in harmony with the demands of the consensus ethic.

By the same token, other behaviours have been relinquished from the social ethic to the personal ethic during the same time span. Sexual behaviour among consenting adults, e.g. homosexuality, once regulated by the social ethic (probably largely for religious reasons), has been relinquished to individuals' personal ethical choices.

Plato has taught us a very valuable lesson about effecting ethical change. If one wishes to change another person's, or society's, ethical beliefs, it is much better to *remind* than to *teach* or, in my martial arts metaphor, to use judo rather than sumo. In other words, if you and I disagree ethically on some matter, it is far better for me to show you that what I am trying to convince you of is already implicit, albeit unnoticed, in what you already believe. Similarly, we cannot force others to believe as we do (sumo); we can, however, show them that their own assumptions, if thought through, lead to a conclusion different from what they currently entertain.

These points are well exemplified in 20th century US history. Prohibition was sumo, not judo, an attempt forcefully to impose a new ethic about drinking on the majority by the minority. As such, it was doomed to fail and, in fact, people drank *more* during Prohibition. Contrast this with Lyndon Johnson's Civil Rights legislation. As himself a Southern redneck, Johnson realized that even Southern rednecks would acquiesce to the following two propositions:

- all humans should be treated equally; and
- black people were human.

Society had just had never bothered to draw the relevant conclusion. If President Johnson had been wrong about this point, if 'writing this large' in the law had not 'reminded' people, civil rights would have been as ineffective as Prohibition!

In sum then, social ethics changes and can be changed by rational means and concerns. With all of this in mind, we can turn back to Gaskell's results about biotechnology, and attempt to elucidate the ethical concerns that led Europeans to reject the creation of transgenic models of human disease despite the acknowledged benefits of such work, be the benefits economic or the augmentation of human health.

Ethics and transgenic animals

As I have pointed out elsewhere (Rollin, 1995), there are two legitimate categories of ethical concern that grow out of the production of transgenic animals. The first set of issues are concerns of safety and risk growing out of the creation of such animals – possible risks to humans, other animals and the environment. However, Gaskell's data indicate clearly that the 'moral unacceptability' of transgenic animal production is logically separate from such risks. In fact, four of the aforementioned biotechnologies, crop plants, food production, transgenic research animals and xenotransplantation, are believed to contain risks. Food production is perceived as harbouring considerably more risk than the production of transgenic research animals, yet only the production of transgenic animals and xenotransplantation is seen as morally unacceptable. (Further, the production of transgenic research animals is seen as 'more useful' than biotechnology in food production, yet is still deemed morally unacceptable.) Thus, we may fairly conclude that while there are certainly real and perceived risks associated with creating transgenic animals for biomedical research, it is not the risks that drive people to consider it morally unacceptable, and thus the moral issues must be sought elsewhere.

Besides risk, there is only one legitimate, as opposed to spurious, moral issue associated with the production of transgenic animals, and that is the question of the well-being of the animals so generated (Rollin, 1995). Indeed, as a purely moral issue, animal welfare is far more vexatious than safety. After all, even a hypothetical researcher with no moral concern about safety would have a prudential interest in assuring the safety of transgenic work, as researchers themselves working with dangerous organisms are certainly *prima facie* more at risk than are members of the general population. (Recall that the world's last smallpox death occurred in the context of laboratory research, and again, that the first European deaths from Marburg virus were laboratory workers.) In addition, any major breach of safety eventuating in catastrophe will almost certainly ramify in truncation of transgenic research, both by engendering restrictive regulation and by virtue of curtailment of funding.

Animal welfare concerns, on the other hand, represent a far greater moral challenge, for concern about animal welfare often does not coincide with perceived self-interest and indeed can exact significant costs in the form of money, time, profit, extra personnel, delay in research, etc. In other words, many researchers traditionally have not equated concern for animal welfare with self-interest, and are thus unlikely to do the right thing for reasons of prudence. Somewhat mitigating this blanket statement is the relatively recent acknowledgement of the fact that failure to assure animal welfare can skew variables relevant to research and actually compromise research (Rollin and Kesel, 1989). Pain, for example, is a significant physiological stressor (Rollin, 1997) but, none the less, the coincidence of the two is far from perfect, and, as we shall shortly see, certain aspects of transgenic animal research do represent an area where welfare could be ignored without obviously jeopardizing the work in question. Thus, moral concerns must take up the slack left after prudential considerations are exhausted.

In my view, the European rejection of the creation of transgenic animals for biomedical research follows from burgeoning international concern with animal welfare, specifically with minimizing animal pain and suffering. As I shall argue shortly, the treatment of animals, traditionally largely left to people's personal ethics, is ever increasingly being grabbed up by the social ethic for reasons we shall detail.

Legislation

There can be no question in the mind of anyone who tracks social change in the western world that concern for animal treatment has become a burgeoning international issue during roughly the past three decades, and particularly since the early 1980s. For example, according to the National Cattlemen's Association, Congress has consistently during this period received more letters, telephone calls and other contacts on issues of animal welfare than on any other issue. The amount of legislation pertaining to animal welfare has proliferated all over the world, with the scope of such laws ranging from the use of animals in zoos, circuses and other entertainment to animal agriculture, animal research, hunting, trapping, teaching, etc. Probably most pervasive initially has been the proliferation of laws, regulations and policies pertaining to the use of animals in biomedical research, with major laws being passed in the USA, Australia, New Zealand, Great Britain, Germany, Sweden and elsewhere, sometimes, as in the USA, against vigorous opposition from the research community, who claimed that any constraints on animal use in research would irrevocably damage human health. So powerful was this opposition that when I, as a principal architect of current US legislation, was asked by the press to predict when our bill would pass, I said '2010'. In fact, so deep was public moral concern that *two* variations on our bill were passed in 1985. A recent article from Europe boldly affirms that 'in nearly every parliament of the Member States of the Council of Europe, there is growing concern for the welfare of laboratory animals' (De Greeve and De Leeuw, 1997). It is correlatively indubitable that the core of all recent legislation and regulation pertaining to animal research in western Europe, North America, Australia and New Zealand is the control and minimization of pain and suffering, as well as an ever-increasing tendency to press forward alternatives to painful animal use. Thus, for example, a January 1998 article in *Laboratory Animals*, (Shalev, 1998) indicated that

> increasing concern within and without the scientific community over pain and distress in animals has made the production of monoclonal antibodies [MAbs] highly controversial ... [with] some European countries having gone as far as banning *in vivo* production of MAbs using the ascites method

In the USA, pain engendered in laboratory animals must be controlled by anaesthesia, analgesia, sedation and early end points, e.g. for tumour growth and disease processes, aimed at minimizing suffering. In Britain, an animal suffering uncontrollable pain and distress must be euthanized as soon as the situation is understood (O'Donoghue, 1992). For reasons of controlling pain and suffering, US journals are increasingly unlikely to publish papers using death as an end point, even though the late end point may well provide valuable information. In other words, globally, there is a consensus emerging that not every human benefit is worth any amount of animal suffering (cf. public rejection of cosmetics companies utilizing safety testing on animals, and the spectacular growth of those companies disavowing such testing).

Experimental animals

Everything we have said thus far is patent, undeniable and clearly points to the profound and worldwide social–ethical concerns about invasive animal use. The message to researchers is thus clear – minimize animal suffering. We may thus conclude that the ethical reason why Europeans reject the creation of genetically engineered animal models for biomedical research is that the creation of such animals is likely to involve severe and uncontrollable pain and suffering. Specifically, transgenic technology now allows us in principle to replicate any human genetic disease in animals. In the past, animal models for human genetic diseases were accomplished by finding adventitious mutations which were propagated through selective breeding. This new technology allows us to create, not to discover, *every* sort of seriously genetically defective animal designed to model human disease. The problem can be focused by a brief quote from a standard book on transgenic animals.

> There are over 3,000 known genetic diseases. The medical costs as well as the social and emotional costs of genetic disease are enormous. Monogenic diseases account for 10% of all admissions to pediatric hospitals in North America ... and 8.5% of all pediatric deaths.... They affect 1% of all live born infants ... and they cause 7% of stillbirths and neonatal deaths.... Those survivors with genetic diseases frequently have significant physical, developmental or social impairment.... At present, medical intervention provides complete relief in only about 12% of Mendelian single-gene diseases; in nearly half of all cases, attempts at therapy provide no help at all. (Karson, 1991)

This is the context in which one needs to think about the animal welfare issues growing out of the possibility of creating transgenic animals used in biomedical research. On the one hand, it is clear that researchers will embrace the creation of animal models of human genetic disease as soon as it is technically feasible to do so. Such models, which introduce the defective human genetic machinery into the animal genome, appear to researchers to provide convenient, inexpensive and, most importantly, high-fidelity models for the study of the gruesome panoply of human genetic diseases outlined in the over 3000 pages of text comprising the sixth edition of the standard work on genetic disease, *The Metabolic Basis of Inherited Disease* (Scriver *et al.*, 1989). Such 'high-fidelity models' may occasionally reduce the numbers of animals used in research, a major consideration for animal welfare, but are more likely to increase the numbers as more researchers engage in hitherto impossible animal research. On the other hand, the creation of such animals can generate inestimable amounts of pain and suffering for these animals, since genetic diseases often involve symptoms of great severity.

Indeed, as I predicted in 1985, the very first attempt to create a model for human genetic disease by embryonic stem cell technology was a mouse designed to replicate Lesch–Nyhans syndrome, a truly hideous disease of xanthine metabolism, recently highlighted in Richard Reston's best-selling *The Cobra Event*, and resulting in compulsive self-mutilation, including children biting off their lips and fingers (Hooper *et al.*, 1987; Keuhn *et al.*, 1987). To the surprise of the researchers, the animals created to model the disease were phenotypically normal, but eventually a symptomatic animal will be possible. The European response, in essence, says no to such chronically suffering animals on moral grounds, regardless of potential benefits.

I would argue that the European ethical rejection of creating such animals is a special case of the emerging social ethic for animals we alluded to earlier, and provides us with an excellent example of what consumers, i.e. members of society, will not accept in animal use, regardless of benefits produced, economic or other, such as improving our ability to do health research. Any enterprise, business or industry using animals must therefore understand the details of this emerging ethic lest it come up short when it is assessed by it. Economic benefit will no longer trump animal welfare considerations as dictated by the new animal consensus ethic. It is to the nature of this ethic we now turn.

THE CHANGING PATTERN OF WESTERN SOCIETY

For as long as we have had articulated ethics in society, we have had a very minimal, limited, consensus ethic (and laws mirroring that ethic) regarding animal treatment. That ethic has been an ethic forbidding *cruelty*, i.e. deliberate, sadistic, useless, wanton, deviant, unnecessary infliction of pain and suffering or wanton neglect upon animals. The Bible condemns this, a number of ancient Greek philosophers condemned it, Catholic theology condemned it, and all civilized societies have laws against it, not only for the sake of the animals, but also because it has long been known that those who are cruel to animals often graduate to people. However, for whatever reason proscriptions against cruelty were promulgated, they generally sufficed as the consensus social ethic about animal treatment until the last few decades. Why has this change occurred? Why are people demanding that our consensus ethic move beyond cruelty?

Vast changes in farming and livestock production

There are a variety of reasons why this has occurred roughly since the 1960s. Most important has been the historically precipitous change in the nature of animal use which occurred approximately at mid-century. The end of World War II witnessed the emergence of two major patterns with profound implications for the traditional social ethic for animals. The first occurred in the area of animal use in biomedical research (Rowan, 1984). From 1900 to 1920, the number of animals used in such research was both low and constant. After 1920, the growth rate increased somewhat and then increased precipitously directly after World War II, when large amounts of money were pumped into research and drug production. Such activity reached a peak in the 1960s. The second pattern occurred in agriculture and grew out of the industrialization of animal agriculture. Between World War II and the mid-1970s, agricultural productivity – including animal products – increased dramatically. In the 100 years between 1820 and 1920, agricultural productivity doubled. After that, productivity continued to double in much shorter and ever-decreasing time periods. The next doubling took 30 years (between 1920 and 1950); the subsequent doubling took 15 years (1950–1965); the next one took only 10 years (1965–1975). As R.E. Taylor points out, the most dramatic change took place after World War II, when productivity increased more than fivefold in 30 years (Taylor, 1992). Fewer workers were producing far more food. Just before World War II, 24% of the US population was involved in production agriculture (Meij, 1960), today the figure is well under 2% (1.7%). Whereas in 1940 each farm worker supplied food for 11 persons in the general population, by 1990 each farm worker was supplying 80 persons. At the same time, the proportion of disposable income spent on food dropped significantly, from 30% in 1950 to 11.8% in 1990 (Taylor, 1992).

There is thus no question that industrialized agriculture, including animal agriculture, is responsible for greatly increased productivity. It is equally clear that the husbandry associated with traditional agriculture has changed significantly as a result of industrialization. Symbolically, Departments of Animal Husbandry in universities in the USA have changed their names to Departments of Animal Science, thereby marking an essential feature of the trend.

For our purposes, several aspects of technological agriculture must be noted. In the first place, as just mentioned, the number of workers has declined significantly, yet the number of animals produced has increased. This has been possible because of mechanization, technological advancement and the consequent capability of confining large numbers of animals in highly capitalized facilities. Of necessity, less attention is paid to individual animals. Second, technological innovations have allowed us to alter the environments in which animals are kept. Whereas in traditional agriculture animals had to be kept in environments for which they had evolved, we can now keep them in environments that are contrary to their natures but congenial to increased productivity. Battery cages for laying hens and gestation crates for sows provide examples of this point. The friction thus engendered is controlled by technology. For example, crowding of poultry would once have been impossible because of flock decimation by disease; now antibiotics and vaccines allow producers to avoid this self-destructive consequence.

New technologies, new problems, new perspectives

A moment's reflection on the development of large-scale animal research and high-technology agriculture elucidates why these innovations have led to the demand for a new ethic for animals in society. In a nutshell, this new technology represents a radically different playing field of animal use from the one that characterized most of human history; in the modern world of agriculture and animal research, the traditional ethic grows increasingly less applicable. A thought experiment makes this clear. Imagine a pie chart that represents all the suffering that animals experience at human hands today. What percentage of that suffering is a result of intentional cruelty of the sort condemned by the anticruelty ethic and laws? When I ask my audiences this question – whether scientists, agriculturalists, animal advocates or members of the general public – I always get the same response: only a fraction of 1%. Few people have ever witnessed overt, intentional cruelty, which thankfully is rare.

Animals for research and teaching

On the other hand, people realize that biomedical and other scientific research, toxicological safety testing, uses of animals in teaching, pharmaceutical product extraction from animals, and so on, all produce far more suffering than does overt cruelty. This suffering comes from creating disease, burns, trauma, fractures and the like in animals in order to study them; producing pain, fear, learned helplessness, aggression and other states for research; poisoning animals to study toxicity; and performing surgery on animals to develop new operative procedures. In addition, suffering is engendered by the housing of research animals. Indeed, a prominent member of the biomedical research community has argued that the discomfort and suffering that animals used in research experience by virtue of being housed under conditions that are convenient for us but inimical to their biological natures, e.g. keeping rodents, which are nocturnal burrowing creatures in polycarbonate cages under artificial, full-time light, far exceed the suffering produced by invasive research protocols (T. Wolfle, personal communication).

Now it is clear that researchers are not intentionally cruel – they are motivated by plausible and decent intentions: to cure disease, advance knowledge, ensure product safety, augment their résumés. None the less, they may inflict great amounts of suffering on the animals they use. (This is not, of course, to suggest that *all* animal research involves pain and suffering.) Furthermore, the traditional ethic of anticruelty and the laws expressing it had no vocabulary for labelling such suffering, since researchers were not maliciously intending to hurt the animals. Indeed, this is eloquently marked by the fact that the cruelty laws exempt animal use in science from their purview. Those who first recognized this suffering as a concern (by and large the humane societies), lacking any vocabulary to describe it, often called researchers cruel, but such a description was clearly inadequate and served only to shut down dialogue between such concerned people and the research community. A new set of concepts beyond cruelty and kindness was needed to discuss the issues associated with burgeoning research animal use.

Intensive livestock production

Precisely the same point is true regarding criticism of confinement, industrialized agriculture. As we shall see, society eventually became aware that new kinds of suffering were engendered by this sort of agriculture. Once again, producers could not be categorized as cruel, yet they were responsible for new types of animal suffering on at least three fronts:

1. Production diseases arise from the new ways in which the animals are produced. For example, liver abscesses in cattle are a function of certain animals' responses to the high-concentrate, low-roughage diet that characterizes feedlot production (this is, of course, not the only cause of liver abscesses). Although a certain percentage of the animals become sick and die, the overall economic efficiency of feedlots is maximized by the provision of such a diet.

2. The huge scale of industrialized agricultural operations and the small profit margin per animal militates against the sort of individual attention that typified much of traditional agriculture. A case that speaks on this point was sent to me by a veterinarian for commentary in the column that I write for the *Canadian Veterinary Journal*:

> You (as a veterinarian) are called to a 500-sow farrow-to-finish swine
> operation to examine a problem with vaginal discharge in sows. There are
> three full-time employees and one manager overseeing approximately five
> thousand animals. As you examine several sows in the crated gestation unit,
> you notice one with a hind leg at an unusual angle and inquire about her
> status. You are told, 'She broke her leg yesterday and she's due to farrow next
> week. We'll let her farrow in here and then we'll shoot her and foster off her
> pigs'. Is it ethically correct to leave the sow with a broken leg for one week
> while you await her farrowing? (Rollin, 1991)

Before commenting on the case, I spoke to the veterinarian who had experienced this incident, a swine practitioner. He explained that such operations run on tiny profit margins and minimal labour. Thus, even when he offered to splint the leg at cost, he was told that the operation could not afford the manpower entailed by separating this sow and caring for her! At this point, he said, he realized that confinement agriculture had gone too far. He had been brought up on a family pig farm, where the animals had names and were provided individual husbandry, and the injured animal would have been treated or, if not, euthanized immediately. 'If it is not feasible to do this in a confinement operation', he said, 'there is something wrong with confinement operations!'

3. Physical and psychological deprivation for animals. The final new source of suffering in industrialized agriculture results from confinement: lack of space, lack of companionship for social animals, inability to move freely, boredom, austerity of environments, and so on. Since the animals evolved for adaptation to extensive environments but are now placed in truncated environments, such deprivation is inevitable. This was not a problem in traditional, extensive agriculture.

These sources of suffering, like the ones in research, are again not captured by the vocabulary of cruelty, nor are they proscribed or even acknowledged by the laws of the anticruelty ethic. Furthermore, they typically do not arise under the traditional agriculture and ethic of husbandry. The development of large-scale uses of animals in science and the (roughly) contemporaneous increase in intensive agriculture engendered significant amounts of new suffering for animals which could not be conceptually encompassed or even discussed in terms of the traditional social ethic proscribing cruelty. At the same time, as public awareness of this suffering increased, the concern for its alleviation and mitigation grew exponentially. Thus the need for a new ethic and a new set of ethical concepts adequate to these technological innovations was created.

Husbandry

To recapitulate: the overwhelming use of animals in society has always been, and still is, agriculture – food, fibre, locomotion and power. The essence of traditional agriculture was *husbandry* (from the Old Norse word for 'bonded to the household'). Husbandry meant putting the animals into the ideal environment they were evolved for, and then augmenting their natural ability to survive with protection from famine, drought, predation, disease, etc. We put square pegs into square holes, round pegs into round holes and created as little friction as possible doing so. If we harmed the animals we harmed ourselves. So powerfully is this 'ancient contract' ingrained in the human psyche, that when the Psalmist wishes to metaphorize God's relationship to Man, he uses a paradigm case of husbandry, the shepherd: 'the Lord is my shepherd, I shall not want... (*Psalm 23*).' Thus, as long as husbandry was the guiding principle of agriculture, the only social ethic needed was prohibition of cruelty, to catch the few deviates who caused suffering for no reason. With the rise of agricultural technology, what I call technological sanders, we could force square pegs into round holes, round pegs into triangular holes and still make profit, because the friction resulted in animal suffering not in diminution of profit. The demand for new ethical categories to deal with suffering that is not caused by cruelty is manifest.

New moral notions

Thus society is faced with the need for new moral categories and laws that reflect those categories in order to deal with animal use in science and agriculture and to limit the animal suffering with which it is increasingly concerned. At the same time, society has gone through almost 50 years of extending its moral categories for *humans* to people who were morally ignored or invisible. As noted earlier, new and viable ethics[1] does not emerge *ex nihilo*. So a plausible and obvious move is for society to continue in its tendency and attempt to extend the moral machinery it has developed for dealing with people, appropriately modified, to animals. This is precisely what has occurred. Society has taken elements of the moral categories it uses for assessing the treatment of people and is in the process of modifying these concepts to make them appropriate for dealing with new issues in the treatment of animals, especially their use in science and confinement agriculture.

What aspect of our ethic for people is being so extended? One that is, in fact, quite applicable to animal use. All human communities face a fundamental problem of weighing the interests of the individual against those of the general welfare. Different societies have provided different answers to this problem. Totalitarian societies opt to devote little concern to the individual, favouring instead the state, or whatever their version of the general welfare may be. At the other extreme, anarchical groups such as communes give primacy to the individual and very little concern to the group – hence they tend to enjoy only transient existence. In our society, however, a balance is struck. Although most of our decisions are made to the benefit of the general welfare, fences are built around individuals to protect their fundamental interests from being sacrificed to the majority. Thus we protect individuals from being silenced even if the majority disapproves of what they say; we protect individuals from having their property seized without recompense even if such seizure benefits the general welfare; we protect individuals from torture even if they have planted a bomb in an elementary school and refuse to divulge its location. We protect those interests of the individual that we consider essential to being human, to *human nature*, from being submerged, even by the common good. Those moral/legal fences that so protect the individual human are called *rights* and are based on plausible assumptions regarding what is essential to being human.

It is this concept to which society in general is looking in order to generate the new moral notions necessary to talk about the treatment of animals in today's world, where cruelty is not the major problem but where such laudable, general welfare goals as efficiency, productivity, knowledge, medical progress and product safety are responsible for the vast majority of animal suffering. People are seeking to 'build fences' around animals to protect them and their interests and natures from being totally submerged for the sake of the general welfare, and are trying to accomplish this goal by going to the legislature.

New ethics for animals

It is necessary to stress here certain things that this ethic, in its mainstream version, is *not* and does not attempt to be. As a mainstream movement, it does not try to give human rights to animals. Since animals do not have the same natures and interests flowing from these natures as humans do, human rights do not fit animals. Animals do not have basic natures that demand speech, religion or property; thus according them these rights would be absurd. On the other hand, animals have natures of their own (what I have, following Aristotle, called their *telos*) (Rollin, 1992) and interests that flow from these natures, and the thwarting of these interests matters to animals as much as the thwarting of speech matters to humans. The agenda is not, for mainstream society, making animals 'equal' to people. It is rather preserving the common sense insight that 'fish gotta swim and birds gotta fly', and suffer if they do not.

Nor is this ethic, in the minds of mainstream society, an abolitionist one, dictating that animals cannot be used by humans. Rather, it is an attempt to constrain *how* they can be used, so as to limit their pain and suffering. In this regard, as a 1993 *Beef Today* article points out (Suther, 1993), the thrust for protection of animal natures is not at all radical; it is very conservative, 'asking for the same sort of husbandry that characterized the overwhelming majority of animal use during all of human history, save the last fifty or so years'. It is not opposed to animal use; it is opposed to animal use that goes against the animals' natures and tries to force square pegs into round holes, leading to friction and suffering. If animals are to be used for food and labour, they should, as they traditionally did, live lives that respect their natures. If animals are to be used to probe nature and cure disease for human benefit, they should not suffer in the process. Thus this new ethic is conservative, not radical, harking back to the animal use that necessitated and thus entailed respect for the animals' natures. It is based on the insight that what we do to animals *matters* to them, just as what we do to humans matters to them, and that consequently we should respect that mattering in our treatment and use of animals as we do in our treatment and use of humans. Since respect for animal nature is no longer automatic as it was in traditional agriculture, society is demanding that it be encoded in law.

Granted, there are activists who do not wish to see animals used in any way by humans, and in the eyes of many animal users, the activists *are* the 'animal rights people'. Yet to focus on them is to eclipse the main point of the animal rights thrust in society in general – it is an effort to constrain *how* we use animals, not an attempt to stop all animal use. Indeed, it is only in the context of animal use that constraints on use make any sense at all! Thus the new mainstream ethic is not an ethic of abolition; it is an effort to reaffirm that the interests of the animals count for themselves, not only in terms of how they benefit us. Like all rights ethics, it accepts that some benefits to be gained by unbridled exploitation will be lost and that there is a cost to protecting the animals' natures. In agriculture, for example, the cost may be higher food prices. However, as the Federation of European Veterinarians asserted, that is a small price for a society to pay to ensure proper treatment of objects of moral concern.

Thus, the new animal rights ethic we have described in society in general should not be viewed as radically different from concerns about animal welfare, as agriculturalists often mistakenly do. It is, in fact the *form* that welfare concerns are taking in the face of what has occurred in science and agriculture since World War II. The demand for rights fills the gap left by the loss of traditional husbandry agriculture and its built-in guarantee of protection of fundamental animal interests.

CONCLUSION

Let us recapitulate our discussion. We have argued that, despite business' rather cavalier ignoring of social ethics, social ethics drives citizen's choices in a very significant way. For example, we saw that the rejection in Europe of one aspect of biotechnology, the creation of transgenic animals for research, is driven by ethical concern for animal suffering, despite the significant economic and health benefits that could accrue to society through such research. We have further argued that concern for animal treatment and suffering is emerging as a major component in our changing social consensus ethic. What does all of this entail for those engaged in animal-based food production?

In the first place, despite business' belief that consumers are driven only by demands for cheap food, this is emphatically not the case. Laws in Sweden, Britain and generally in the European community have forced producers in those countries to back off from industrialized confinement agriculture, despite its high profitability and provision of cheap, plentiful food. While people certainly want cheap and plentiful food, they also want animals to live decent lives, as they did under husbandry. Furthermore, social ethics all over the world has begun to express concern about other hidden costs of non-husbandry, efficiency and productivity-oriented industrialized animal agriculture, air and water pollution, dissolution of rural communities, loss of independent small farmers to large corporate domination of agriculture, and issues of food safety growing out of such 'efficiency-driven' practices as feeding dead animal tissue to cattle, which was the cause of the bovine spongiform encephalopathy (BSE) outbreak in Britain. Surveys repeatedly indicate that the public does care a great deal about how farm animals live their lives, even though they are destined to die in the end; does care that their natures are respected; and does believe emphatically that animals' lives matter to them.

Thus it behoves agribusiness to begin thinking about the ethics (or lack thereof) underlying confinement animal agriculture, especially in the USA. To begin thinking about if and how husbandry can be restored to an industrialized agriculture. To begin examining meeting those needs of animals which, when thwarted, do not affect profit. To begin attending to demonstrated public concerns in other countries. To think twice before zealously exporting industrialized animal agriculture systems to developing countries. Even if these businesses are motivated *strictly* by self-interest, it is extremely unwise prudentially to ignore the pulse of social ethics. What sense is there, after all, in capitalizing billions of dollars in confinement pig production units if society will inevitably shut them down for both animal welfare and environmental reasons?

It is obviously utopian to believe that we can return totally to family-run, pastoral agriculture. However, that does not mean that we are stuck with the opposite extreme. Somewhere within the constraints imposed by ever-increasing urbanization, burgeoning population and the need for reasonably priced food lies the possibility of a moral agriculture, treating animals fairly, land and water wisely, and agricultural communities as precious renewable resources. The ideal of husbandry and the new consumer ethic for animals we have described provides a beacon lighting the morally defensible path.

REFERENCES

Gaskell, G. and the European Public Concerted Action Group (1997) Europe ambivalent on biotechnology. *Nature* 387, 845pp.

De Greeve, P. and de Leeuw, W. (1997) Developments in alternatives and animal use in Europe. In: Van Zutphen, L. and Balls, M. (eds), *Animal Alternatives: Welfare and Ethics*. Elsevier, Amsterdam.

Hooper, M., Hardy, K., Handyside, A., Hunter, S. and Monk, M. (1987) HPRT-deficient (Lesch–Nyhan) mouse embryos derived from germline colonization by cultured cells. *Nature* 326, 292ff.

Karson, E. (1991) Principles of gene transfer and the treatment of disease. In: First, N.

and Haseltine, F. (eds), *Transgenic Animals.* Butterworth-Heinemann, Boston, Massachusetts.

Kuehn, M.R., Bradley, A., Robertson, E.J. and Evans, M.J. (1987) A potential model for Lesch–Nyhan syndrome through introduction of HPRT mutations into mice. *Nature* 326, 295ff.

Meij, S.L. (1960) *Mechanization in Agriculture.* Quadrangle Books, Chicago, Illinois.

O'Donoghue, P. (1992) European regulation of animal experiments. *Laboratory Animals* September, 1991, p. 20ff.

Rollin, B.E. (1991) An ethicist's commentary on the case of the sow with a broken leg waiting to farrow. *Canadian Veterinary Journal* 32.

Rollin, B.E. (1992) *Animal Rights and Human Morality,* 2nd Edn. Prometheus Books. Buffalo, New York.

Rollin, B.E. (1995) *The Frankenstein Syndrome: Ethical and Social Issues in the Genetic Engineering of Animals.* Cambridge University Press, New York.

Rollin, B.E. (1997) Pain and ideology in human and veterinary medicine. *Seminars in Veterinary Medicine and Surgery – Small Animal: Pain* 12; 56–61.

Rollin, B.E. (1998) *The Unheeded Cry: Animal Consciousness, Animal Pain and Science,* 2nd edn. Iowa State University Press, Ames, Iowa.

Rollin, B.E. and Kesel, M.L. (eds) (1989) *The Experimental Animal in Biomedical Research,* Vol. 1. CRC Press, Boca Raton, Florida.

Rowan, A. (1984) *Of Mice, Models, and Men.* SUNY Press, Albany, New York.

Scriver, C., Beaudet, W., Sly, W. and Vale, D. (eds) (1989) *The Metabolic Basis of Inherited Disease,* Vols I and II. McGraw-Hill, New York.

Shalev, M. (1998) European and US regulation of monoclonal antibodies. *Laboratory Animals* January, p. 15ff.

Suther, S. (1993) Are you an animal rightist? *Beef Today,* April.

Taylor, R.E. (1992) *Scientific Farm Animal Production.* Macmillan, New York.

An Alternative Ethic for Animals

Gary L. Comstock

Bioethics Program, Iowa State University, Ames, Iowa, USA

INTRODUCTION

What are the ethical foundations of modern animal science, and are they philosophically justifiable? Let us begin at the beginning.

WHAT IS ETHICS?

What makes an issue in animal science a case of ethics as opposed to a case of custom, law, religion or science? A few words about what ethics is not.

Ethics is not custom

A very useful introduction to these matters is found in Rachels (1993). There are things we do as a matter of habit that are not necessarily right. Taxi drivers in some cities in the USA customarily give blank receipts to their fares on the understanding that the person will inflate the price paid, receive a higher amount in reimbursement from their company, and pass a bit along to the taxi driver. The fact that drivers and customers act in this way does not make it right for them to act this way. There are also things that are not customary that are not necessarily immoral. We do not usually explain the sometimes sordid details of our divorce proceedings to strangers who casually ask how we are doing, but it would not be immoral to offer them such information. We should not confuse morally justifiable acts with habits or customs.

Ethics is not law

Some things are legal but clearly immoral, such as psychologically abusing your spouse, and some things were at one time illegal but were not necessarily immoral, such as an African-American sitting in the front of a bus in Alabama in 1950. Ethical rules are not identical with the laws of a community, even though communities rightly strive to form their laws in accordance with what is ethically justifiable.

Ethics is not, strictly put, religion

Historically, the ethical values of many cultures have evolved from religious traditions, and the faith traditions have often been our primary repositories, incubators and champions of virtue and character. However, philosophically, what is right or good is not necessarily identical with what a particular religion teaches. There is the obvious problem that some religions teach prejudice and discrimination, but there is a deeper problem, which Plato pointed out long ago: whatever it is that makes something good is not that God commands it. God commands something, rather, because it is good. God commands us, for example, not to starve our children to death not because God is capricious and happens to decide at the moment that murdering children is distasteful. Rather, murdering children is wrong, and God, being omniscient, knows that it is wrong. Being omnibenevolent as well, God is good, and commands us not to do what is wrong. The important point is that things that are wrong are wrong whether or not God forbids them. To put it another way, ethics is independent of God's will.

There are good reasons to separate public policy decisions and the revelations of particular faiths, and not only because religious people disagree among themselves about what is right. Countries that try to separate church matters from matters of state attempt to make regulations and laws not on the basis of sacred truths revealed to a few, but on the basis of broader, secular, principles upon which people from diverse religious backgrounds – and no religious background – can agree. It is possible to reach a consensus about moral issues without invoking religious authorities. Consider one example. In the USA, many people once believed that it was morally wrong to allow women to vote. Some traditions thought it imperative on biblical and theological grounds to keep women out of the public sphere, whereas other traditions supported the suffrage movement on grounds that were equally theological and biblical. However, once the culture removed the issue from the sphere of religion and looked at the facts about women, it could not justify its view that women should not vote. The general population came to a consensus that the policy should be changed because justice demanded it. There was no need to settle the vexing theological questions; the question was settled, and in the right way, on non-religious grounds.

Ethics is not science

Ethical judgements must always be based on good science, but science is not ethics. Science is a descriptive discipline aimed at explanation and then prediction. Using science, we try to discover and articulate natural laws and regularities that govern the behaviour and relationships of objects in the natural world. Ethics is a normative discipline aimed at prescribing conduct in which we try to discover and articulate moral laws that ought to govern human behaviour. There are scientific questions we can answer without having to think about ethics at all, just as there are arithmetical questions we can answer without having to do ethics. However, there are few ethical questions one can resolve without basing one's judgements on accurate scientific information.

Ethics is not naïve relativism

Sometimes when we use the word ethics, we mean the variety of loosely knit sets of rules that implicitly guide the conduct of different groups. On this interpretation, there is a wide variety of ethics. Whether a group systematically articulates its principles or not, it has a set of action guiding rules which, for the purposes of clarity, we may call the group's morality. Different cultures, then, have different moralities; indeed, it is the differences in morality that constitute the most dramatic differences between groups. Ancient Hebrew and contemporary Jewish traditions encourage the husbandry and selective breeding of animals as part of the theological mandate by God to subdue the earth. This culture also, however, proscribes the eating of molluscs and the consumption of blood. Jains in India, on the other hand, prohibit the eating of molluscs, the consumption of blood and the selective breeding of animals on the grounds that all of life is sacred and animals and humans are linked through the cycle of karma and reincarnation. On the other hand, Christian cultures throughout history have allowed not only the breeding of animals and eating of shellfish, but the consumption of blood as well.

If ethics meant nothing more than morality, then our task in a chapter devoted to animal ethics would be an empirical, descriptive, project: to survey and articulate the various moral codes regarding human use of animals for food and fibre. I am not aware of any systematic code of ethics set forth by an animal science professional organization. The American Veterinary Medicine Association has a code that sets forth principles of professional conduct, but it makes no attempt to justify these rules philosophically, nor to provide a theoretical basis for the fundamental assumption that animals exist to serve human interests. An ethical theory must explore and question such assumptions.

However, one might object that even if there are no formal statements of the ethical code of animal science, there certainly are implicit sets of principles that guide conduct in this area: that is true. The problem, then – as the Jewish, Jain and Christian example suggests – is that there are many such sets of principles. Does it not follow from the fact of diversity and conflict that there is no truth to the matter? Well, if there was no truth in ethics, then there would be no truth in cultural relativism either, because it would be logically impossible for one culture to disagree with another about practical moral issues. If, say, the wanton killing of innocent humans was not permissible in my culture but was in yours and there were no truth of the matter, then there would be no grounds for me to argue that it is wrong for you to kill and eat me. So far so good for naïve cultural relativism. However, some cultures hold that there is a truth to the matter, and they believe that wanton killing is wrong everywhere and every place. How does a relativist honour that culture's morality? So much the worse for cultural relativism. The mere fact that cultures disagree about moral matters does not suggest that relativism is true.

Some of my students, eager to adopt relativism because of moral diversity, do not want to do the work of seriously engaging the various cultures, examining their arguments and subjecting their theories to critical scrutiny. To try to articulate a generalized set of rules, rules that apply to anyone at any time in any place who is facing a set of circumstances identical to those faced by someone else, is a difficult chore. It is also fraught with the terrible temptations and dangers of cultural imperialism. However, that chore, with all of its attendant problems, is what I will mean henceforth by ethics. Ethics will not, in other words, tell George that it is acceptable to kill a cow in circumstances q, r and s, while telling Jorge that it is unacceptable to kill a cow in exactly the same set of circumstances.

If ethics is not science, law, custom, religion or naïve relativism, what is it?

Applied ethics is a normative discipline, aimed ultimately at prescribing and governing. Ethicists try to discover and articulate moral laws that ought to govern human conduct and choices. The difference here is that whereas science tries to tell us what is, applied ethics tries to tell us what ought to be. The sphere of ethics is the sphere of moral choices, and moral choices differ from the operations of the natural world in that moral agents can choose to act in ways that science cannot predict, in ways that are contrary to our instincts and nature's physical laws. None of the moral questions facing animal science may be answered by simply doing more scientific research, no matter how carefully crafted. It is true that defensible moral judgements must be based on good science, and no ethical argument is sound if it makes claims that are empirically false. None the less, normative arguments require general moral principles as premises, and figuring out whether these principles are justified requires philosophical reflection about our ethical theories and our shared values.

When we discuss animal welfare and environmental ethics, we invariably have different views about the matter, and we invariably invoke arguments, explicitly or implicitly, to support our conclusions. Therefore, the study of animal ethics is always, at least in part, the study of moral arguments, which is to say the study of premises, conclusions and the validity of moving from premises a, b and c, to conclusion d. A better definition of ethics, then, is the study of arguments about what is morally right and wrong, good and bad. By an ethic I will mean a clear, non-contradictory, comprehensive and generalized set of rules intended to govern human behaviour. By ethics, I will mean the study of arguments about these theories, arguments about what things are good and bad, and which actions right and wrong.

WHAT IS ANIMAL ETHICS?

Drawing on the previous discussion, we may now define animal ethics as the study of arguments about what things are good and bad, and which actions right and wrong, in the use of animals for food and fibre. We also may define an animal ethic as a clear, non-contradictory, comprehensive and generalized set of rules intended to govern human behaviour in the use of animals for food and fibre. Notice that an animal ethic is a theory with practical consequences. It is a theory insofar as it is a clear and coherent set of rules: a set, in other words, of non-contradictory propositions. It has practical consequences insofar as these rules are intended to guide action. As a practical theory, it is capable of resolving conflicts between humans and animals.

Conflicts relating to animal production now facing the world community

- As billions of people around the world seek increasingly to emulate the high meat-consuming diets of the developed countries, will the globe's natural resource base be able to sustain an industrial agricultural system devoted to high-volume, low-cost, monocultural production of animal feedstuffs? How much land can 10 billion meat eaters spare for wildlife (Waggoner, 1994)?
- As demand for meat increases, will animal scientists be asked to genetically engineer animals that will be 'happy' living in close confinement conditions their ancestors would have found intolerable (Rollin, 1995, pp. 192–193)?
- As demand for meat increases, how likely is it that livestock producers

will become resented free riders when it comes to paying the costs of environmental externalities such as soil erosion and loss of biodiversity in rangeland?

- As global trade intensifies, are peasant animal farmers likely to be compensated fairly for their intellectual property in 'upstream' contributions to animal breeding? Is the playing field level for smallholders in the South who must compete against multinational corporations in the North, corporations which patent the intellectual property advances of animal biotechnology, thereby capturing almost all of the 'downstream' revenues?

- Will the control of life become increasingly concentrated in the hands of a few corporations in the North, leaving the malnourished of the South destined for lives of persistent poverty and foreclosed opportunity?

- How lamentable is it that some large-scale pig producers in developed countries seem increasingly to be acting as bad neighbours, fouling rural air, polluting streams with animal offal and contributing to the demise of independently owned and operated farms? Examples are quoted by Thompson (1997).

- Is it fair if most of the research at state-supported Animal Science Departments in Colleges of Agriculture in the USA is aimed at benefiting high-volume, low-margin, single commodity producers rather than low-volume, speciality crop, organic and low-input mixed farmers?

- Should the public be concerned that animal scientists, who have now cloned mammals, may be preparing the way for the cloning of humans as well?

- How concerned should animal scientists be that their profession seems to some observers to be part of the cause of these problems, rather than part of the solution? For information on training animal scientists to play a key role in dealing with ethical issues, see Iowa State University Bioethics Institute (1998).

WHAT ARE THE ETHICAL FOUNDATIONS OF ANIMAL SCIENCE?

What I mean by ethical foundations are the metaphysical beliefs, ethical principles and moral rules, implicit or explicit, that serve to justify the practices and institutions of a group. Metaphysical beliefs are ontological commitments, beliefs about the way the universe is structured; Jews and Moslems believe that the world was created by one God who gave humans dominion over animals and divided the animals into clean and unclean species. Jains, on the other hand, believe that all living things are interdependent parts of one another and that all species, including insects, are sacred. Moral principles are ethical propositions that follow from one's metaphysical commitments. Moslems believe that certain animals are created to be sacrificed to Allah; Jains believe that all animals are created to live out their normal life spans. Moral rules, finally, are normative claims about the way humans ought to act. Moslems hold that one's religious duties on hajj, the pilgrimage to Mecca, include the obligation ritually to slaughter a goat. Jains believe that all animal slaughter is wrong, and follow rules requiring them rigorously to avoid stepping on insects. These are ethical foundations for competing animal sciences.

The ethical foundations of a branch of science, therefore, are the metaphysical beliefs, ethical principles and moral rules that serve to justify the practices and institutions of that scientific field. The best way to articulate the ethical foundations of contemporary animal science would be to trace its historical development. At the risk of grossly oversimplifying an extremely long and complex story, let me quickly describe two historical precursors of animal science: hunters and husbands; then I address the position of scientists.

HUNTERS

The religious and ritualistic aspects of the primal hunter–gatherer are well documented, but we rarely acknowledge that these people have an animal science with its own ethical foundations. The foundations are not articulated in systematic theories but in the form of myths and legends that serve in turn to guide character and shape attitudes. The sacred narratives of hunting–gathering cultures depict animals as totems, dangerous beings of spiritual power.

Hunters often regarded themselves as standing in a spiritual relationship with animals, as noted by Krech (1994, p. 16).

> The Assiniboine of the 1790s, according to a North West Company trader, offered a pipe to an old bull or all the buffalo in a pen and said something like, 'My Grandfather, we are glad to see you, and happy to find that you are not come to us in a shameful manner, for you have brought plenty of your young men with you. Be not angry at us; we are obliged to destroy you to make ourselves live'. After this speech, and after all smoked their pipes, the animals were killed and vermilion coloured swansdown placed on the head of each, after which each person was 'at liberty to take what he thinks proper'.

Also, hunters typically regard the taking of animal life as a religious undertaking (Krech, 1994, p. 16): 'Smoking tobacco and offering the pipe to propitiate whoever had power to ensure success were common. Failure in the hunt, if not due to an impetuous hunter, was easily ascribed to improperly performed ritual'.

Metaphysical principles underlying hunting

The metaphysical principles that support the animal science of subsistence hunting, therefore, include the beliefs that: each animal has a spiritual consciousness and individual identity of its own; the gods will favour those who perform the proper rituals; and that while humans should exist in balance with other species, humans are none the less entitled to favour themselves over other species.

The hunter's ethical principles include the ideas that one ought to act in harmony with the natural world; and that one must respect the sacrality and power of individual animals.

Several moral rules follow from these metaphysical foundations, including two specific items: man is justified in killing animals to preserve his own life; and, man is entitled to kill and eat animals so long as he preserves adequate herd size for future use.

The strength of the hunter's ethical foundation is its recognition that game animals have their own social lives and mental states, their own unique individuality. The weaknesses are its faulty scientific theories as, for example, its belief in spontaneous generation. Nor is it a strength of the hunter's animal science to believe that religious ritual is the primary means of ensuring success in feeding oneself. As husbandmen came to realize, selective breeding is superior to incantation as a means of guaranteeing food supply.

HUSBANDS

Pastoral nomads and farmers engaged in selective breeding and husbandry of animals. Their metaphysical principles often include the belief that animals are made, owned and cared for by God, who has put humans in charge of them here on earth. Animals are also thought to have a telos or ends of their own, and these ends establish certain moral boundaries on what the shepherd and farmer and rancher can do to and with them. The husbandry ethic is not unfamiliar. As a youth, I worked as a hired hand at Circle 23 ranch in Buena Vista, Colorado. The ranch ran about 200 head of cattle on 140 acres of irrigated pasture. We were up at 6 a.m. to carry oats to horses in the corral near the barn, had breakfast at 7.30 a.m., moved irrigation pipe until noon, had lunch, and then moved irrigation pipe and fixed fences all afternoon. For cows to survive Colorado winters, humans must work long hard hours during the summer.

In spring, we moved cows to summer grazing lands in the national forest. We went for weeks without seeing cattle. None the less, almost all of our activities were devoted to sustaining animals, even when they were out of sight. We spent months moving irrigation pipe in July to make hay in September to feed cattle in January. At other times, we were closely involved with animals, as when a cow had trouble in delivery. Once, unable to pull a breech birth out by its feet, we tied a rope around the calf's hooves and pulled it with a tractor. When the cow dragged along the barn floor we chained her to a post to separate her from the calf. My children think the procedure sounds inhumane, but we did it to save the life of the cow and her calf. Horses were always fed before breakfast; cows had constant access to fresh water and salt; balanced rations were provided scrupulously to all; and a jar of milky penicillin was stored in the refrigerator to treat infections.

Metaphysical principles underlying husbandry

The metaphysical principles that support the animal science of husbandry, therefore, include the beliefs that: animals are placed on earth by God in order to serve human needs; that each animal has a place in the agro-ecosystem household; that God helps those who help themselves; and that while humans should exist in balance with other species, humans are none the less entitled to favour themselves over other species.

The husband's ethical principles includes the idea that one ought to treat animals humanely, caring for their needs and slaughtering them painlessly. Several moral rules follow from these foundations, including: one is justified in breeding animals in order to guarantee one's source of food; and one is entitled to kill and eat animals to improve one's living standards so long as one does not cause them unnecessary pain or suffering.

The strength of the husband's ethical foundation is its recognition that animals are sentient, that humane care of animals is critical, and that humans have evolved with animals as part of an extended kin network. Animals are subordinates, but co-members none the less of the wider household. The weaknesses are its ignorance of contemporary virology, pathology, physiology and pharmacology. Nor is it a strength of the husband's animal science to have lost the hunter's recognition of the individuality and uniqueness of each animal.

SCIENTISTS

Since the beginning of husbandry agriculture some 10,000 years ago, domesticated animals and humans have enjoyed a mutually beneficial relationship. Humans benefited from the food and companionship animals provided, and animals benefited from humane care. With the change from the husbandry to the animal science model in this century, however, came new tools, and intensified pressure, to select animals for desirable traits. We quickly learned that animals make powerful experimental tools, and it is no surprise that we now have come to accept the view that we should care for animals in the same way that we care for laboratory instruments. As long as they still suit our purposes and fulfil a useful function for us, we care for them. When they become damaged, out of date or come to the end of the life span we appoint for them, we discard them, get new ones and start over again. Farm animals therefore benefit under the animal science paradigm insofar as they are in a position to serve our interests. Many of them live far better lives than their ancestors lived because of the extraordinary gains we have made in scientific knowledge: in, for example, diagnostics, pathology, microbiology, virology, immunology and physiology. By the same token, many of them live far shorter and more deprived lives than their ancestors lived because of the extraordinary gains we have made in raising animals quickly and efficiently to market weight under intensive confinement conditions.

Metaphysical principles underlying animal science

The central metaphysical principle of animal science is this: all and only human beings have moral standing, i.e. we can draw a line that will protect all humans as subjects while simultaneously excluding all animals from such protection. Animals are more like clocks than children, having no interests of their own to constrain our treatment of them. Animals serve as our laboratories, instruments fit for exploration to advance the state of knowledge. If animals are not sentient, then there are no moral boundaries on the amount of pain, suffering or deprivation that we can cause them because they are incapable of feeling any pain, suffering or deprivation. The only laws that seem to constrain animal science, therefore, are those drawn by the professional's desire to avoid alarming 'the emotional (i.e. unscientific) public'.

The ethical principles that guide scientific animal breeding are utilitarian; will a certain experiment or procedure lead to the greater good for the greater number of humans? Theological convictions, not being scientific, have no place in modern animal science, and the hunter's belief that each animal has a unique identity and spiritual power is abandoned.

Transgenic animals: a recent example showing the assumptions of science

In 1982, scientists injected a rat growth hormone gene into the chromosome of a mouse. The resultant animal grew to twice its parents' size. In 1985, researchers at the US Department of Agriculture inserted a human growth hormone gene into the chromosome of a pig. At the Agricultural Experiment Station in Beltsville, Maryland, experimenters successfully microinjected the piece of DNA encoding the production of human somatotropin into the nucleus of a fertilized pig egg. The extracted embryo was re-implanted into a sow's uterus, the pregnant animal came to term, and the first piglet in history with a human gene was born.

Nineteen transgenic swine lived through birth and into maturity. Several expressed elevated levels of the growth hormone gene, but none grew more quickly or to greater size than their counterparts in the control group. However, many suffered from 'deleterious pleiotropic effects', medical problems not afflicting the controls (Fox, 1989). Those animals developed abnormally and exhibited deformed bodies and skulls. Some had swollen legs; others had ulcers, crossed eyes, renal disease or arthritis. Of 29 founder pigs, 19 expressed either human growth hormone or bovine growth hormone. Among those exhibiting long-term elevated levels of bovine growth hormone, health was generally poor. Many seemed to suffer from decreased immune function and were susceptible to pneumonia (Fox, 1989). All were sterile. Later, Pursel *et al.* (1989) said: 'the pigs had a high incidence of gastric ulcers, arthritis, cardiomegaly, dermatitis, and renal disease'. They concluded that if transgenic swine were to be produced as successfully as transgenic mice, 'better control of transgene expression, a different genetic background, or a modified husbandry regimen' would be required.

In preparing this manuscript, I asked a dozen of the most prominent transgenic farm animal (TFA) researchers two questions:

1. What is the total number of TFAs now being produced annually?
2. What percentage of TFAs produced annually are significantly worse off than they would have been had they not been tampered with at the embryonic stage?

Results of my informal survey showed that there is no central clearing-house for the information sought in question 1. Therefore, calculating the percentage sought in question 2 is impossible. However, of the researchers who hazarded guesses, all agreed that the number of transgenic cattle could be probably counted on one's fingers; that the number of founder pigs would be less than 200, and that goats and sheep would be in the thousands. While I did not provide them with a clear definition of what 'being significantly worse off' means, none of them seemed to think that the concept was fuzzy or difficult to understand. They all agreed that the percentage which are 'significantly worse off' is probably very low. Indeed, while all of the reporters were cautious, most seemed to think that the Beltsville pigs were the only transgenic animals that suffered as a result of genetic engineering. Their reasons seemed to fall into one of three categories.

1. Embryos do not suffer. The result of physical damage to the embryo is almost always prenatal lethality, so few transgenic large animal embryos are actually brought to term. Results of transgene effect may be construct dependent or integration site dependent. Lines with deleterious phenotypes as a result of a transgene integration site are almost always discarded,

unless the phenotype is of scientific interest in its own right. This has
happened occasionally in mice but apparently has not been reported in
larger animals. There is little point in producing an unhealthy farm animal
because only healthy productive farm animals will make the farmer any
money.

2. Cost. Whereas the expense involved in making transgenic mice is relatively low, the cost of livestock is high. Most of the work to develop animal
models of disease occurs in mice because much work involves gene knock-
out experiments, and the stem cell lines needed for these experiments
currently are only available for mice. TFA researchers, therefore, discard
large animal embryonic constructs if they know that the constructs are
analogous to transgenic mouse constructs that have led to problems in mice
when brought to term. Farm animals are not used as models for specific
gene function, so many of the conditions that have been seen in mice are
not likely to be seen in TFAs, and the number of TFAs at risk is kept to a
minimum.

3. Frequency of attempts. Government agencies are reticent to fund large
animal transgenics since the 1985 USDA Beltsville growth hormone in pigs
studies gave such bad phenotypes. They generally seek answers from transgenic mice, as mentioned above, if the researchers with large farm animals
do not first provide it. Also, private corporations are even less likely to take
unnecessary financial risks than government.

A word of caution is appropriate. Much transgenic animal research is
now occurring in private industry laboratories where goats and sheep are
being used to produce altered proteins in milk. Information about these
animals is proprietary and not freely shared. Therefore, there may be inci-
dents of transgenic animal suffering that only a few people know about. We
simply do not know.

Transgenic animals: a further example

Thus, it appears that after the problematic Beltsville pig incident, transgenic farm animal production has been relatively free of animal welfare
problems. However, the assumption persists that animals are here for our
use. It is most troubling in the case of rodents, where we have suddenly
come to accept the novel idea that we are entitled to make animals much
worse off in order to benefit humans. An experiment performed at the
European Molecular Biology Laboratory in Heidelberg, Germany, and
reported in *Nature*, in which experimenters, trying to identify the gene for
limb growth, found it, knocked it out and brought into the world 23 female
and 48 male genetically engineered mice with foreshortened limbs (Zakany
and Duboule, 1993). What could be worse for the mice so afflicted? The
details follow:

In the forelimbs, the radius and the distal epicondyle of the humerus were absent. The bony mass was reduced distally and incomplete sets of carpals, metacarpals and phalanges were observed. In hindlimbs ... digit I, its metatarsal and tarsal supporting bones were lost as well as digit II. In addition, distal phalanges were missing. In both fore and hind-limbs, proximal–distal fusions were often observed. Distal truncations of tibia, fibula and ulna occurred, with fusion between fibula and talus, resulting in a remodelling of the ankle.

We read of the experiment at the University of Texas that knocked out a gene (*Lim*1), and resulted in the birth of four headless mice that could not breathe (Oliwestein, 1996; Krauthammer, 1998). How could such experiments be conceived, much less performed, if society did not widely accept the view that animals may be exploited for virtually any reason someone deems 'scientific'?

Effects of the ethical foundations of animal science

What have been the consequences for farm animals of the ethical foundations of animal science? As consumers increasingly demanded a standardized product, animal scientists seem to have become unconcerned with the interests of the animals except insofar as those interests coincided with the breeder's interests. The breeders' interests are almost exclusively economic, with some surprising results. The wild sheep, capable of producing about 1 kg of thick rough wool each year as protective insulation, has been made over into a virtual wool machine, producing some 20 kg of fine downy wool for sweaters each year.

What happened to the animal in the process? Whereas sheep naturally shed almost all of their wool each spring during their seasonal moulting period, intensively bred animals have lost most of their biorhythms and do not moult with seasonal regularity. The sheep of modern animal science must be shorn by humans (Goude, 1981, p. 67). Wild cattle that once produced a few hundred millilitres of milk each year have now been made over into virtual milk machines capable of producing 15,000 litres (Goude, 1981). We now breed food animals that cannot perform the biological functions characteristic of their species, such as turkeys that cannot fly and cows that will not care for their calves. Modern animal science has gone so far in changing the genomes and phenotypes of our food animals that philosophers assert, somewhat grandiosely, that we have created these 'artefacts', that the animals are more like machines than like wild animals (Callicott, 1989; Taylor, 1986).

As happens with inbreeding among humans, narrowing gene pools often brings unintended results. When companion animals are backcrossed for anatomical features consumers consider desirable, the animals often suffer problems, such as respiratory difficulties, anatomical abnormalities or sensory deprivation. The dog has perhaps been treated worst of all as we have selected for traits that render some dogs virtually blind, lame or incapable of breathing. We have bred in great numbers dogs that seem to loathe themselves as much as they hate others.

The story does not end with breeding, however. Because the rationalization of agriculture required low cost and high volume, we sought methods by which to house food animals in closer and closer confinement. Raising the number of animals per space increases the numbers a farmer can take to market. There are obvious limits. When too many animals are crowded together, living conditions become so stressful that animals engage in behaviours that decrease their numbers. Loss of food animals is not only economically harmful to farmers; it is bad for animals. Pigs bite off each others' tails; chickens resort to cannibalism and self-destructive pecking; veal calves suffer muscle atrophy and anaemia.

Are the ethical foundations of animal science justifiable?

I have argued that modern animal science rests on the view that we can draw a line to protect all humans as subjects with full moral standing while excluding all animals from such protections. However, can we justify such a line? (Author's note: the following seven paragraphs are reprinted from my commentary (Comstock, 1998), with the kind permission of the editor).

The history of theology and philosophy knows several attempts to do so, and the most powerful attempts hold either that only humans have souls; or that only humans are rational; or that only humans have moral rights. Alan Donagan (1977) identifies each of these candidates in discussing the basis of informed consent laws. According to Donagan, the main reasons usually offered as a basis for respecting humans are:

1. In nature, as it is known to us, human beings have a dignity and worth that is unique;
2. The Kantian principle that a human being is never to be used merely as a means, but always at the same time as an end; and
3. A principle identified and articulated in the Declaration of Independence ... that every human being is endowed with an inalienable right to life, liberty, and the pursuit of happiness (Donagan, 1977, p. 162).

Let us examine these reasons beginning with the second. The second principle is the Kantian principle that humans are not to be used as means to an end because humans are rational creatures, able to reflect on their ends and choose their own ends for themselves. However, this is a high bar, and many humans do not measure up to it. None of the following humans is rational: a fetus in the ninth month; a neonate; the severely mentally enfeebled; an aged patient in advanced stages of Alzheimer's disease; all humans in irreversibly comatose states. These so-called 'marginal humans' are not rational human beings, but they constitute a group of individuals that morality probably ought to protect. However, they will not be protected if only the rational among us are legitimate objects of moral concern. Donagan's second principle, therefore, sets the mark too high.

Consequently, Donagan's first principle is attractive. It suggests that humans have special 'dignity and worth' because they are unique. They are the only members of the human species. This principle avoids setting the standard too high. It is very inclusive, and will protect infants and the mentally enfeebled.

However, there are two problems with this way of justifying the protection of innocents. First, there may be other innocent beings that we should protect, and yet this principle offers us no basis on which to protect them. Suppose there are other beings in the universe of whom we presently do not know, beings that are at least as caring and rational as we are. Think of a science fiction example, of ET, the extraterrestrial. If ET landed tomorrow, should we not have in place a system of morality that would extend moral protection to him? However, if we hold that membership in the human species is a necessary requirement in order to be protected morally, then we could not protect ET without radically changing our beliefs. Indeed, we would be justified in killing ET for the same sort of trivial reasons we kill stray ants in our kitchen; the creature 'just bugs us', it is 'out of place', it 'does not belong in the house'. However, it would be brutish to kill a being as sensitive and compassionate as ET simply because his unusual shape rubs you up the wrong way, or because you want to look at his insides or because you would like to know how his cooked flesh might taste. The first problem with Donagan's first principle is that it fails to protect all innocents.

Why, you might respond, should we worry about trying to protect all innocent individuals? Not *everyone* shares the intuition that we should protect all innocents. Peter Singer (1980a), for example, denies that 'gross mental defectives' have a right to life equal to an ordinary adult's right to life. He holds this view because of his anti-speciesist belief that 'merely being a member of the species *Homo sapiens* cannot carry with it any special moral status'. Singer's position is a radical one; he is unsure whether his position can protect the mentally enfeebled, and he seems inclined to believe that we must be ready to change our attitudes toward marginal humans:

> [My position] involves holding that mental defectives do not have a right to life, and therefore might be killed for food – if we should develop a taste for human flesh – or (and this really might appeal to some people) for the purpose of scientific experimentation. (Singer, 1980a)

Singer provides two other possible positions, but rejects the first one because he does not believe, as Tom Regan (1983) does, that all humans have an equal right to life. He is doubtful about the third position, that marginal humans and animals have some kind of serious claim to life – whether we call it a 'right' does not matter much – by virtue of which, while we ought not to take their lives except for very weighty reasons, they do not have as strict a right to life as do persons. In accordance with this view, we might hold, for instance, that it is wrong to kill either mentally defective humans or animals for food if an alternative diet is available, but not wrong to do so if the only alternative is starvation.

Singer hesitates to adopt this view because he is not sure that the replaceability argument can be met. That argument holds that lives are replaceable, and the pleasures that one human or animal may experience may be replaced by the pleasures of another human or animal should the first individual be killed. I believe this argument can be met by seeing that modified forms of utilitarianism will not ultimately protect innocents, but I do not have room to make the case here. This may be compared with Pluhar (1990).

However, insofar as a society has resources to care for mental defectives, it ought to do so simply as a way of demonstrating its humanity. I do not disagree with Singer that 'gross mental defectives' may lack a moral right to life, assuming that the individuals in question truly lack sentience, consciousness and all cognitive life. It may be that truly grossly mental defectives who live in perpetual and irreversibly comatose states, or who lack a brain, thereby lack even the most basic of moral rights. Where Singer goes wrong, however, is in thinking that moral rights are the only fences protecting such individuals from being killed for trivial reasons. Even grossly defective humans may be morally valuable for reasons that cannot be captured in the language of moral rights, so I disagree with Singer that the proper response to the marginal human problem is to revise common morality so as to exclude marginal humans from the protections morality offers. The proper response is to figure out what it is that all but the most grossly marginal humans possess. Once we figure that out, then we have identified the characteristic that allows them to be included in morality's protections. Then the right thing to do is to extend morality's protections to all other individuals who possess the minimal characteristic that marginal humans possess. I do not have the space here to justify my answer, but my suspicion is that the right characteristic is 'desires' (see Singer, 1980b; Regan, 1983; Varner, 1998). Do farm animals have desires?

The second problem with holding that all and only human beings are unique is that it begs the question. It fails to explain why being human is sufficient to merit special protections not afforded to others. In order to justify a difference in moral treatment between species, there must be some relevant moral difference, some capacity, X, which only humans possess, which no non-humans possess and which succeeds in entitling us to preferential treatment. What might X be? Theologians traditionally have thought of X as 'a soul' or 'God's image', and attributed our sacredness to that unique characteristic. The problem here is that many religious people disagree that only humans have souls. Hindus, for example, hold that animals have souls and, within Judaism and Christianity, many agree. The Catholic saint, Francis of Assisi, held that animals have souls, as do C.S. Lewis and Andrew Linzey, recent Anglican theologians. So there is not a consensus in Christianity, much less among the world's religions, that animals lack 'X', where X is 'a soul'.

There are secular ways of trying to identify a unique human quality. Kantian and contractarian philosophers think of free will, or autonomy, as the characteristic that entitles us to be treated as ends in ourselves. Autonomy is defined in various ways, but many of the following human characteristics play central roles in the arguments of Kantians: our ability to make choices, to possess free will and to exercise our will so as to pursue our own conception of the good life without infringing on the ability of others freely to pursue their conception of the good life. The Kantian view is that all and only humans are autonomous agents, and on this basis we are uniquely entitled to be treated as ends and never as means only. The contractarian view is that all and only humans who can make and keep promises are entitled to be part of morality's protections.

The marginal human problem plagues both the Kantian and contractarian responses too, however. Neither the Kantian nor the contractarian can protect marginal humans, humans who are neither rational nor moral agents. Is the way open, therefore, to using marginal humans in scientific experiments? Clearly not, because they have desires, needs and feelings. Systems of morality that not only fail to protect the frail and weak among us, but which consciously exclude them as objects of legitimate moral concern, should hardly qualify as an ethical system at all. Perhaps morality should not protect the very grossly marginal, such as the anencephalic infant who does not have, and will never have, desires or a unique psychological identity of its own. However, anyone who is able to take an interest in something is a moral patient.

Arguably, morality should protect all moral patients. Elsewhere I have argued that being autonomous is not a necessary characteristic for having moral standing because a good many adults turn out not to be autonomous, according to one definition of autonomous (see Comstock, 1992).

Verdict on the moral principle of modern animal science

Based upon the foregoing discussion, it is evident that the fundamental moral principle of modern animal science does not protect all moral patients; it does not protect marginal humans and it does not protect sentient animals. Therefore, the ethical foundation of animal science – the idea that all humans have, and all animals lack, moral standing – is seriously flawed.

CONCLUSION

Modern animal science does not encourage us to view animals as fellow citizens of the biosphere. It turns a blind eye to their interests and, in the process, hides from us the fact that we may be quietly disconnecting ourselves from the natural world. The conjunction of the following three facts is not mere happenstance.

1. Agriculture is becoming increasingly geographically widespread, rationalized and industrialized.
2. Wild animal species are disappearing.
3. Farm animal genetic stock is becoming increasingly narrow.

The convergence of these trends suggests that our attitude to nature and animals is increasingly reductionistic. Now I have some suggestions. Let us take a lesson from history and recover the hunter's appreciation for the individuality of each animal. Let us feel justified in killing animals for food when it is absolutely necessary for self-preservation, but not when we can easily meet our dietary needs without slaughter. Let us recover the husband's consideration for the humane treatment of animals in our care, and reintegrate animals with our agrohouseholds and agroecosystems. Finally, for anyone desiring to continue with unchanged animal science foundations, we must ask: how do you answer the question raised here? The question is: 'what is the justification for the belief that an animal is worth less than a human of comparable mental capacities?'.

Practical actions for animal scientists

Finally, following the arguments presented in this chapter, we can identify two practical implications and options for animal scientists. First, animal scientists should undertake a global interdisciplinary research effort to develop truly humane animal production systems. Such systems would have three features. They will:

- be integrated with local environments, households and economies;
- permit the animals in the system to live out their normal life spans and to pursue the desires and form the social groups that their ancestors have pursued and formed for generations;
- kill animals for meat only when the life of the animal was no longer worth living to the animal.

Second, animal scientists should work to develop nutritious, tasty and aesthetically satisfying meat substitutes derived either from animal by-products – milk, eggs, blood gathered from animals reared free of the stress of intensive confinement and isolation – or from plant biomass.

The next phase in the history of animal science promises to be very exciting. The discipline is poised to advance beyond its modernist, philosophically untenable, principles to a very different, but also in some ways strangely familiar post-modernist set of values. It is heartening that this book is being published and contributing to a sensitive dialogue seeking to evaluate these trends.

SUMMARY

What are the ethical foundations of modern animal science, and are they philosophically justifiable? I begin by distinguishing ethics from custom, law, religion and science, and then explain what I mean by the idea of a scientific discipline having ethical foundations. I argue that the ethical foundation of modern animal science is the conviction that all, and only, humans have moral standing, and that all non-human animals lack moral standing. I present reasons leading to the conclusion that the ethical foundation of modern animal science is seriously flawed, and end with some brief thoughts about post-modern animal science.

REFERENCES

Callicott, B. (1989) *In Defense of the Land Ethic: Essays in Environmental Philosophy.* SUNY Press, Albany, New York.

Comstock, G. (1992) The moral irrelevance of autonomy. *Between the Species* 8, 15–27.

Comstock, G. (1998) Research with transgenic animals: obligations and issues. *The Journal of BioLaw and Business* 2, 51–54.

Donagan, A. (1977) Informed consent in therapy and experimentation. *The Journal of Medicine and Philosophy* 2, 310–327, partially reprinted in Gorovitz, S. (ed.) (1983) *Moral Problems in Medicine.* 2nd edn. Prentice-Hall, Englewood Cliffs, New Jersey, citation at p. 162.

Fox, M.W. (1989) Genetic engineering and animal welfare. *Applied Animal Behaviour Science* 22, 107.

Goude, A. (1981) *The Human Impact: Man's Role in Environmental Change.* MIT Press, Cambridge, Massachusetts.

Iowa State University Bioethics Institute (1998). Information on short courses is available from Comstock, 421 Catt Hall, Iowa State University, Ames, IA, 50011-1306, USA. email: comstock@iastate.edu

Krauthammer, C. (1998) Of headless mice . . . and men: the ultimate cloning horror: human organ farms. *Time* 151, January 19.

Krech, S., 3rd (1994) Ecology and the American Indian. *Ideas* 16.

Oliwenstein, L. (1996) Headless. *Discover Magazine* January.

Pluhar, E. (1990) Utilitarian killing, replacement, and rights, *Journal of Agricultural Ethics* 3, 147–171.

Pursel, V.G., Pinkert, C.A., Miller, K.F., Bolt, D.J., Campbell, R.G., Palmiter, R.D., Brinster, R.L. and Hammer, R.E. (1989) Genetic engineering of livestock. *Science* 244, 1281.

Rachels, J. (1993) *The Elements of Moral Philosophy*, 2nd Edn. McGraw-Hill, New York.

Regan, T. (1983) *The Case for Animal Rights.* University of California Press, Berkeley, CA.

Rollin, B.E. (1995) *The Frankenstein Syndrome: Ethical and Social Issues in the Genetic Engineering of Animals.* Cambridge University Press, Cambridge.

Singer, P. (1980a) Animals and the value of Life. In: Regan, T. (ed.), *Matters of Life and Death: New Introductory Essays in Moral Philosophy.* Random House, New York, pp. 280–321.

Singer, P. (1980b) Utilitarianism and vegetarianism. *Philosophy and Public Affairs 9.*

Taylor, P. (1986) Respect for Nature: *a Theory of Environmental Ethics.* Princeton University Press, Princeton, New Jersey.

Thompson, P. (1997) From a philosopher's perspective: how should animal scientists meet the challenge of contentious issues? Unpublished paper, pages 15–16.

Varner, G. (1998) *In Nature's Interests? Interests, Animal Rights and Environmental Ethics.* Oxford University Press, New York.

Waggoner, P. (1994) How much land can ten billion people spare for nature? *Council for Agricultural Science and Technology Task Force Report* No. 121. Ames.

Zakany, J. and Duboule, D. (1993) Correlation of expression of Wnt-1 in developing limbs with abnormalities in growth and skeletal patterning. *Nature* 362, 546–549.

Consumer Expectations for Animal Products: Availability, Price, Safety and Quality

6

Shin-haeng Huh[*]

Korea Consumer Protection Board, Korea

INTRODUCTION

In general, consumers expect that, as economic conditions change, animal products will change. However, economic conditions are linked directly or indirectly with other parts of the world. In oriental philosophy, there is a saying that 'the whole universe is one body'. This statement implies that everything is linked together like the organs in a human body. In a similar way, therefore, consumer expectations are closely linked with other changes, including economic growth, living standards, trade patterns, the demand for and supply of animal products, health problems, and so on. There are five areas which need consideration in order to gain a better understanding of the consumer position in changing conditions.

First, an analysis of consumer expectations is helpful to gain an understanding of the trade patterns in the global arena of animal products. Second, it is essential to look briefly at technological improvements and their effect upon the availability of animal products. Third, it is often helpful to examine price trends of animal products. Fourth, there is the growing public concern about the safety of animal products. Finally, there is the issue of consumer expectations in terms of the quality of animal products.

GLOBALIZATION AND FREE TRADE

The world is changing rapidly as a result of the revolutions taking place globally. Actually, three interdependent revolutions are occurring

*Present address: Seoul Agricultural and Marine Products Corporation, 600 Garah-dong, Song pa-ku, Seoul 138–701, Korea.

simultaneously in the following fields: communications, transportation and computer technology. We will consider each separately.

First, the communication revolution now offers a wireless cellular telephone, at a reasonable price which is thus available to anyone. As a result, the new era of 'one person, one phone' is placing all humans into one communication system, like the nervous system in a human body. All the people on earth are becoming one big family through this communications revolution. This development is leading to a world which crosses time zones.

Second, the transportation revolution continues to take place. The first phase of this revolution took place when James Watt invented the steam engine in 1782, leading in a few decades to the railways. The second phase began with the invention of the internal combustion engine in 1886. The third phase started after a jet engine was invented during the Second World War. Now, the fourth phase of this revolution is beginning with the development by the US National Aeronautical and Space Administration (NASA) of hypersonic passenger aircraft. This fourth transportation revolution will have the effect of bringing the whole world together as close as the people in a small village in which people can fly anywhere within a few hours. These changes will lead to a world which knows no physical boundaries.

The third revolution with computers started with the invention of vacuum tubes, then transistors, followed by integrated circuits which then became high-density integrated circuits, and now continues with the possibilities of artificial intelligence. It seems that further improvements in technology for computers will be limitless, and are leading to a cyberspace world in which most human activities are carried out through a small multimedia box, enabling mankind to overcome the limitations of both time and space.

A major result of these three revolutions is to turn the world into a single market which will eventually function like a human body. With this analogy, we can think of communication as the nervous system, transportation as the circulatory system and computers as the brain. We may imagine a man, who initially is in a coma and who can neither eat nor move. After some time, he wakes up and starts making simple movements. However, over time, he gains an improved nervous system, circulatory system and brain and becomes fully functional as a human being. In a similar way, human society is likely to become one borderless single market behaving with the coherence of a normal human body. We have to start thinking of the whole earth, after 4.6 billion years, beginning to behave as a huge single living organism.

The type of change we speak of is called regionalization and globalization. Regional integration today may be found in the European Union (EU), the North American Free Trade Agreement (NAFTA), the Asia Pacific Economic Cooperation (APEC), and so on. These regional groups continue to expand by including additional member countries. The EU was enlarged from 12 to 15 member states in January 1995. The summit meeting of the Americas in December 1994 paved the way for 34 countries in the

Americas to conclude a negotiation for a Free Trade Area of the Americas (FTAA) by the year 2005. We can expect to see more regional integration in the future.

The most important event leading to globalization was the establishment in January 1995 of the World Trade Organization (WTO). Through this body, the nation states are working towards integration of the world into a single market. In an economic sense, the world is becoming a borderless body in which products and services move freely without barriers. Consequently, free trade is becoming inevitable throughout the world.

Free trade increases the volume of trade among countries. An increase in the volume of trade results in more products and more investment. Animal products are not an exception in this world trade, which will result in the expansion of animal production and in increased consumption of animal products in trading countries.

We can already see a steady increase in the import of beef and veal, pork and poultry meat in the world, as shown in Table 6.1 in metric tonnes (t). The world import of beef and veal increased rapidly from 2.6 million metric tonnes (Mt) in 1982 to 4.1 Mt in 1997, global pork imports increased from 877,000 t in 1982 to 2.5 Mt in 1997, poultry imports increased from 1.1 Mt in 1982 to 5.0 Mt in 1997.

Table 6.1. World imports of livestock and poultry meat, 1982–1998 (×1000 t).

Year	Beef/veal[a]	Pork[a]	Poultry meat[b]
1982	2631	877	1146
1983	2619	910	1114
1984	2647	1247	961
1985	2598	1476	1044
1986	2962	1555	1069
1987	2915	1658	1262
1988	3069	1850	1248
1989	3407	1837	1386
1990	3444	1929	1597
1991	3689	1945	1680
1992	3795	1548	1882
1993	3773	1856	2285
1994	4135	2286	2974
1995	4059	2508	3888
1996	3895	2583	4663
1997[c]	4105	2527	5000
1998[d]	4419	2741	5430

[a]Carcass weight equivalent; [b]ready-to-cook equivalent; [c]preliminary; [d]forecast.
Source: USDA, FAS, Livestock and Poultry (1997b); USDA, World Markets and Trade, USDA, October (1997c).

We may see similar increases in the export of these animal products and poultry meat. As we know, imports and exports are the two sides of a coin.

There are many factors affecting the continuing increase in the trade of animal products. Such factors may include the rising population, increasing production and consumption, the growth and development of economies, and so on. However, in any situation, if world markets are closed, then there will be no trade at all. The opening of markets is the most important factor affecting increased trade in animal products.

This is exemplified by the trade in dairy products. Global butter imports increased from 382,000 t in 1983 to 522,000 t in 1997. Global cheese imports also increased from 459,000 t in 1983 to 793,000 t in 1997. The imports of both butter and cheese are estimated to increase by the year 2006 to 807,000 t and 964,000 t respectively, according to the Food and Agricultural Policy Research Institute (1997).

What are the future prospects for trade in animal products? It depends on various factors such as the economic development of the trading countries, technology improvement, health problems, major wars, and so on. However, if there are no disastrous problems then, as more countries open their domestic markets, the volume of trade in animal products will definitely continue to increase in the future. The expectation is that there will be more production and higher consumption of animal products. Therefore, consumers can expect the trade in livestock products to increase, as they enjoy consuming more and better animal products.

TECHNOLOGY IMPROVEMENT AND AVAILABILITY OF ANIMAL PRODUCTS

In general, as its economy develops, a society moves successively from a primitive stage to a farming society, then to industrialization, to an information and/or knowledge society, and so on. Some economists have identified these different societies with stages of economic development.

In a primitive society, people hunted wild animals for food and clothing. When society became more settled, it began cultivation and farming, and people began to domesticate livestock for food along with planting crops and vegetables. There was no special feed for livestock, which were grazed on natural pastures or fed waste products from crops. A family usually raised one or two head of cattle and/or pigs and a few chickens.

Later, in industrial societies, capital and technology became major factors of production, and this was reflected in the livestock industry. Farmers expanded their livestock flocks and herds by using more capital and adopting better technology. As a result, livestock production increased.

As livestock farmers became more affluent, they invested in new facilities and technology such as breeding, feeding, nutrition, health and equipment, leading to an increase in livestock production. As high technology is developed in an information society, more livestock products become available to consumers outside the production system.

We see a clear upward trend in the production of livestock products and poultry meat in Table 6.2 in metric tonnes (t). The world production of beef and veal increased from 40.8 Mt in 1982 to 48.3 Mt in 1997.

Global pork production also increased from 48.6 Mt in 1982 to 80.4 Mt in 1997. Global poultry production increased from 23.0 Mt in 1982 to 53.8 Mt in 1997.

We can expect to see an increasing production trend as further advanced technologies are developed. These new technologies help livestock farmers to overcome limitations such as shortages of land, labour and capital in producing animal products.

As long as we continue to develop new technologies, the availability of livestock products may be infinite. The only problem may be the cost–benefit ratio in developing technologies for the livestock industry. However, as time passes, this problem is likely to be solved, giving consumers a continuously higher availability of animal products.

Table 6.2. World production of livestock and poultry meat, 1982–1998 (×1000 t).

Year	Beef/veal[a]	Pork[a]	Poultry meat[b]
1982	40,822	48,570	23,040
1983	41,121	50,741	23,529
1984	42,411	51,882	24,253
1985	43,562	54,463	26,140
1986	44,353	56,646	27,257
1987	44,894	58,401	29,088
1988	48,217	62,616	32,693
1989	48,439	63,534	33,962
1990	49,056	64,806	35,845
1991	48,921	65,910	37,750
1992	45,697	64,943	37,573
1993	45,554	67,110	40,534
1994	46,940	70,827	43,852
1995	48,010	75,143	47,654
1996	48,605	79,177	50,531
1997[c]	48,290	80,435	53,842
1998[d]	48,031	82,493	56,998

[a]Carcass weight equivalent; [b]ready-to-cook equivalent; [c]preliminary; [d]forecast.
Source: USDA, FAS, Livestock and Poultry (1997b); USDA, World Markets and Trade, USDA, October (1997c).

PURCHASING PRICE OF ANIMAL PRODUCTS

The nominal price of animal products generally continues to increase in most countries. However, the real price of animal products as well as the real income level affect demands for animal products.

For consumers, both nominal and real prices of animal products do not mean much without reference to their income level. Even if the price of animal products is low, poor consumers with low income cannot buy them. On the other hand, in countries with high prices for livestock products, rich consumers with high income can easily purchase them. Therefore, whether a consumer can buy livestock products or not depends largely upon his income level as well as the relative price of the products. The purchasing power is affected not only by the real price of livestock products but also by the disposable income of the consumers. This means that income mainly determines the purchasing power of the consumer.

Has the purchasing price represented by the relative price of livestock products increased or decreased so far? Will it increase or decrease in the future? The general trend of the purchase price has been decreasing for a long time in most advanced countries, as well as major trading countries including newly industrialized countries and some major developing countries, too. Of course, there have been some fluctuations in the movement of price changes depending on unexpected factors such as unusual weather conditions, widespread disease, the long depression of economies, etc. Taking these unusual factors into account, the purchase price of animal products has decreased, and will continue to do so.

If the purchase or the real price of animal products continues to decline, then the demand for them will increase continuously throughout the world, except in the case of some less-developed countries. This will lead to more production and trade in animal products. The result enables consumers to expect more animal protein and a higher standard of living.

SAFETY OF LIVESTOCK PRODUCTS AND CONSUMER PROTECTION

In the past, most people did not pay very much attention to food safety. In Korea, for example, people usually did not know about safety problems and therefore did not give much thought to food safety until a food-related illness prompted concern.

In the past, for example, some rural people might have died suddenly after returning from a wedding or a funeral ceremony. Villagers did not know why some died right after returning from the ceremonies. Some thought they died because they were possessed by a ghost. Others believed they simply had bad luck. However, they had one thing in common. They had eaten a particular food such as pork and/or other livestock products. Whenever rural people had a ceremonial meeting in their villages, they usually butchered a pig and a few chickens to serve to the gathering. On many occasions, the pork was not prepared in a safe manner.

Nowadays, as economies develop, consumer income increases, living standards rise, technology and science to battle diseases improve, so people around the world are becoming more concerned with food safety. Even rural people in Korea are learning about the cause of the death of their neighbours. The common cause of those deaths, of course, is unsafe food and disease.

Even most advanced countries including the USA and many European countries had not paid sufficient attention to the safety of livestock products until new problems occurred. A recent outstanding problem is the disastrous case of mad cow disease in the UK which for several years destroyed the beef export industry.

Another animal disease problem was anthrax in cattle in the state of Victoria in Australia. Foot-and-mouth disease wiped out the whole pig industry in Taiwan in February 1997. Within a very short period in recent years, these three animal diseases have threatened consumers throughout the world. Further, classical swine fever resulted in the destruction of tens of thousands of pigs in The Netherlands in 1997 and influenza virus H5N1 destroyed the chicken industry in Hong Kong in August 1997. In 1999 a new pig virus killed humans in Malaysia.

Threats from animal diseases appear to be real, numerous and varied. In recent years, outbreaks have occurred of *Escherichia coli* O157:H7, *Listeria monocytogenes*, *Campylobacter* and *Salmonella* in meat. Microorganisms are believed to be the main cause of most food-borne illnesses. The estimated morbidity and mortality associated with food-borne illnesses in the USA in 1993 are shown in Table 6.3.

The second source threatening the safety of animal products for human consumption may be drugs that are used to protect animals from, and to treat them for, diseases. These drugs include streptomycin, novobiocin, ampicillin, penicillin, oxytetracycline, chloramphenicol, sulphamerazine, thiabendazole and sulphadimethoxine. In addition, there are other drugs and hormones used as additives in feed. These include antibiotics, synthetic animal drugs and growth-promoting hormones, including zeranol and

Table 6.3. Estimated human morbidity and mortality (numbers of people) associated with food-borne illness in the USA, 1993.

Causative agent (% food-borne)	Morbidity	Mortality
Salmonella (87–96)	696,000–3,840,000	696–3,840
Campylobacter (55–70)	1,418,494–1,805,356	110–511
E. coli O157:H7 (80)	8,000–16,000	160–400
Listeria monocytogenes (85–95)	1,526–1,767	377–475

Source: Craigmill (1996), 'Food Safety Issues in the Twenty-first Century', International Symposium on Food Hygiene and Safety, 30 May–3 June, Seoul, Korea.

diethylstilbestrol for pigs and calves. If livestock farmers misuse the sub-stances in terms of dose, usage time and methods, they can harm the meat, and damage the human body.

The third source threatening the safety of animal products is pesticides remaining in pasture and feed grains. These potentially dangerous pesti-cides include DDT, dieldrin and aldrin, BHC, endrin and captan.

The fourth source of lack of safety is mycotoxins passing through live-stock animals and finally reaching the human body. Of course, mixed feeds are primarily contaminated with mycotoxin. In the 1960s, thousands of turkeys were killed by so-called 'turkey X disease' from the contamination of peanut feed with *Aspergillus flavus* in the UK.

The fifth source threatening the safety of animal products are radio-active materials and other environmental contaminants. No one can forget the disastrous nuclear accident in which radioactive materials leaked from a nuclear plant in Chernobyl, USSR, on 26 April 1986. Livestock and humans, as well as all other living creatures, have long been affected by radioactive materials. The safety of animal products is also threatened by environmental materials such as mercury, lead, cadmium, arsenic, phenol, tricholoroethylene and polychlorinated biphenyl (PCB). These harmful heavy metals and chemical substances remain in livestock animals. As many countries are becoming industrialized, this problem appears to be a serious and growing concern of animal scientists.

To ensure the safety of food, including animal products, the inter-national Codex Alimentarius Commission adopted Guidelines for the Appli-cation of the Hazard Analysis Critical Control Point (HACCP) System in 1993. The HACCP is a preventive system identifying specific hazards and preventive measures for control to ensure the safety of food. It system-atically estimates the likelihood of a food safety hazard. It also establishes specific control measures at each identified critical control point of pro-duction, from harvesting to processing, and to consumption of the final product. The term critical control point (CCP) is defined as a point, step or procedure in a food process at which control can be applied and, as a result of which, a food safety hazard can be prevented, eliminated or reduced to acceptable levels.

The HACCP system has seven principles as follows:

1. Conduct a hazard analysis;
2. Determine CCP in the process;
3. Establish critical limits for preventive measures associated with each identified CCP and monitor the system;
4. Take corrective action when deviations occur;
5. Establish effective record-keeping procedures that document the HACCP system;
6. Establish procedures for verification that the HACCP system is working correctly.

A growing number of countries have adopted the HACCP system. Australia, New Zealand, the USA, Canada, Japan and Korea are leading countries in adopting the HACCP system.

What can consumers expect to see with regards to the safety of animal products? As time passes, it is not easy to say whether animal products will be safer. However, it seems reasonable to conclude that consumers can expect increasingly safe products as a variety of improved technologies is used to ensure the safety of food. However, some scientists do not agree; they believe that the safety of animal products is likely to become worse as more drugs and hormones are used, and as radioactive materials as well as other environmental contaminants increase with economic development.

There is debate about whether the safety of animal products is increasing or not. One view is that safety is definitely improved as the whole of society develops. However, some scientists believe that sources threatening safety have remained constant; they believe this in spite of technological development to fight diseases and harmful chemicals and materials. As a result of advanced technology, scientists are discovering more diseases and other hazards, and thus most consumers think eating animal products is becoming more dangerous. However, the reality is the opposite, as technology to identify diseases has been improved tremendously. Hence, most experts believe that animal products are becoming increasingly safer.

However, it is important to remember that nothing is guaranteed. As far as the safety of animal products is concerned, no one can be sure. Safety depends largely on the efforts of people developing the livestock industry as well as processing, transportation, storage and cooking industries. As long as all the people involved and the governments make continued efforts to ensure food safety, consumers can rest assured.

CONSUMER EXPECTATIONS ON QUALITY OF ANIMAL PRODUCTS

Here we are not speaking of food safety, but of qualities which are desired and approved by the consumer. There is no doubt that most consumers expect higher quality animal products. The problem is in defining and knowing what constitutes better quality.

Everything is 'relative' in value and judgement. This is also true in the consumption of animal products. For instance, people living in a primitive society might consider fat pork meat to be the best quality. However, people with a high income in a developed country would consider it to be the lowest quality. Therefore, consumers' opinions on quality of animal products may depend upon their income level or a consumption pattern, and consequently upon stages of economic development.

In general, the food consumption pattern is known to change in five consecutive stages as consumers' income or living standards increase, i.e. survival; recognition; choice; preference; art.

In the 'survival' stage, people do not care about quality, they are simply concerned with the quantity of food needed for their survival. In the 'recognition' stage, consumers begin to see the importance of nutrition. In the 'choice' stage, consumers may choose food items based on taste and other criteria. In the 'preference' stage, consumers have a preference in selecting their food and animal products for special or other purposes. In the final stage, 'art', consumers can enjoy having food with natural flavour and other features. Hence, the quality of animal products can be treated differently based on the food consumption pattern.

In the 'art' or final stage, there are a few stipulations for animal products to be of high quality. The high-quality product must be safe and fulfil consumers' needs and preferences. The consumers' preference for a high-quality product could be fulfilled with sensuous, nutritional and processing properties. For a sensuous property, high-quality beef, for instance, should have a good marbling and a tender texture with low cholesterol. On the other hand, the grading system also becomes very important for high-quality meat. Good quality products can be produced and processed, based on the grading system, so that consumers are able to choose what they want in the market.

As economies develop, consumers expect an improvement in the areas of breeding, foodstuff and feeding, processing, hygiene and other related sectors of technology. These improvements may result in animal products of high quality, so that most consumers can expect to consume increasingly better quality animal products.

CONCLUSION

We have reviewed consumer expectations with respect to availability, price, safety and quality of animal products. As one can imagine, consumer expectations are high and they want a safe product of better quality at a low price. These expectations can be fulfilled if economies develop in a stable fashion without war and natural disasters, and continued investment is made for developing the livestock industry, as well as other related industries.

As the global economy is integrated into one in which free trade expands, consumer expectations can easily be met. However, the safety of animal products could be challenged more frequently. Whenever there is an outbreak of animal disease, livestock products are endangered, and consumers all over the world worry about consuming them. Most consumers, however, are well informed by the mass media in this information age, and it is impossible to hide factors related to human health. It is an open society and a society of consumers' sovereignty.

Once the consumer turns his or her back on animal products, livestock producers may lose their jobs. Then the size of the livestock industry would be reduced or, in the worst case, the industry could be destroyed. In order to strengthen the industry, a lot of investment and human effort is needed. In the future, the livestock industry and trade volume may fluctuate in terms of availability and price. Imagine a small lake and a big sea, we can see little waves in the former, but huge billows in the latter. The only difference between them is the volume of water. The same goes for the trade and consumption of animal products. As the volume of livestock production and trade increase, the livestock industry tends to be destabilized by uncertainty. Uncertainty in the livestock industry throughout the world will be a situation we are likely to face in the coming century.

Whether we can have a stable and developing livestock industry to fulfil consumer expectation will be largely dependent on efforts made by animal scientists as well as governments. As long as all related people continue to make efforts to develop and maintain a safe livestock industry, consumers may enjoy products at a reasonable price and with improved safety as well.

REFERENCES

American Meat Export Association (1996) *US Meat*. Summer, Autumn.

Caselli, P. (1997) The single market after 1992: problems and prospects. In: Khosrow, F. (ed.), *International Trade in the 21st Century*. Pergamon, Oxford.

Food and Agricultural Policy Research Institute (1997) *1997 World Agricultural Outlook*. Staff Report No. 27-97, Iowa State University, University of Missouri, Columbia, USA.

Hingley, A. (1997) *Focus on Food Safety*. US Food and Drug Administration. FDA Consumer, September–October.

Institute for Livestock Technology (1996) *New Livestock Technology*. Publication No. 31255-51890-57-9603, Korea.

Kaplan, B., McGuigan A. and Williamson, C.C. (1997) *HACCP for Hamburgers from Farm to Table*. Food Safety Information of the USDA.

Katsutoshi, M. (1996) *Recent Trends of Food Poisoning: Outlook in Japan and Molecular Approaches to Diagnosis of Foodborne Pathogenic Bacteria*. Proceedings of an International Symposium on Food Hygiene and Safety, Seoul, Korea.

Kim, Y.-K. (1997) Production strategies for high quality, Hanwoo beef. In: *The 7th Short Course on Feed Technology, Korean Society of Animal Nutrition and Feedstuffs*. Department of Animal Science, Dong-A University, Pusan, Korea, pp. 306–320.

Lee, J.-M. (1996) *Symposium on HACCP Operation and Marketing of American Fresh Chilled Meat*. Seoul, Korea. Available from Products and Utility Division, National Livestock Research Institute, Rural Development Administration, Suwon, Gyonggi Province, Korea.

Lee, M.-H. (1997) Feed hygiene and safety of livestock products. In: *The 7th Short Course on Feed Technology, Korean Society of Animal Nutrition and Feedstuffs*. College of Veterinary Medicine, Seoul National University, Suwon, Gyonggi Province, Korea, pp. 567–600.

Levy, B. (1997) Globalization and regionalization: main issues in international trade patterns. In: Khosrow, F. (ed.), *International Trade in the 21st Century*. Pergamon, Oxford.

Park, J.-H. (1997) Current situation and suggestion for the improvement of feed safety control policy. In: *The 7th Short Course on Feed Technology, Korean Society of Animal Nutrition and Feedstuffs.* Available from Feed Industry Research Institute, Korea Feed Association, Seoul, Korea, pp. 116–148.

Park, Y.-H. (1997) Foodborne pathogenic microorganisms and HACCP. In: *Symposium on Recent Rapid Detection Method of Hazard Microorganism for Safe Meat Production.* Seoul, Korea. Available from Department of Microbiology College of Veterinary Medicine, Seoul National University, Gyonggi Province, Korea.

USDA Food Safety and Inspection Service (1996a) *Meat and Poultry Hotline, Food Safety Features.* Food Safety Publication.

USDA Food Safety and Inspection Service, FS010410, (1996b) *Adult Resource Material, Preventing Foodborne Illness – a Guide to Safe Food Handling.*

USDA Food Safety and Inspection Service (1997a) *The Food Safety Education Newsletter,* Vol. 2, No. 3. Edited by The Food Safety and Education Staff.

USDA Foreign Agricultural Service of USDA (1997b) *Livestock and Poultry: World Market and Trade.* Circular Series 2–96.

USDA Foreign Agricultural Service on line (1997c) *Dairy: World Markets and Trade.* October.

WAAP-FAO (1995) Supply of livestock products to rapidly expanding urban populations. In: Kim, W.-Y. and Ha, K.-J. (eds), *Proceedings of an International Symposium.* Seoul, Korea.

Ethics, Culture and Development: Livestock, Poverty and Quality of Rural Life

7

Denis Goulet

University of Notre Dame, Notre Dame, Indiana, USA

INTRODUCTION

Two major questions are addressed. First, how do ethics and culture relate to development? This question involves a survey of the functions of ethics and culture in development arenas. The second question is then addressed, namely, what effect does animal husbandry have on poverty and rural life?

DEVELOPMENT ARENAS

The role of ethics

As successive UNDP Human Development Reports (1990, 1992, 1994) note, economic development is the means, human development is the end. Most governments, international agencies, business firms and non-governmental organizations (NGOs), however, mistake the means for the end and pursue economic growth as the end. In so doing, they sacrifice human development. The Canadian economist David Pollock (1980) asks the vital question: 'Does man live by GNP alone?'. Indeed, development decision-making poses in a new form three ancient ethical questions.

- What is the relationship between the fullness of good and the abundance of goods?
- What are the foundations of justice in and among societies?
- What criteria govern the posture of societies toward nature?

The provision of sound normative, institutional and behavioural answers to these questions is what makes a country developed. It follows, therefore, that not every nation with a high per capita income is truly developed. These ancient questions become contemporary developmental questions, and answers to them supplied by ancient wisdom are rendered uncertain, because of the following modern conditions:

- the vast scale of human activities;
- technical complexity and the specialized division of labour which ensues therefrom;
- the web of interdependence which transforms local happenings into global events, and causes international conflicts to impinge on local destinies; and
- the ever-shortening time lag between changes, proposed or imposed, and the deadline human communities face to react in ways which protect their integrity.

Not every 'way of doing ethics' is adequate, however, for the task of integrating development's diagnostic and policy issues into its value realms. It is not enough for ethicists merely to posit morally good ends to economic or political actions. Nor does it suffice for them to pass judgement in the light of moral rules extrinsic to the instrumentalities employed in pursuit of those ends. Rather, ethicists must lay bare the value content of instrumentalities from within their proper dynamism. In addition to gauging the moral merits of competing ends, ethicists must show whether export promotion policies favour equity or not, whether they consolidate fragile local cultures or not. Development ethics conducts a phenomenological 'peeling away' of positive and negative values latent in the means chosen by technical experts and political decision-makers. Those making moral judgements must master technical data pertinent to problems under study in ways which specialists recognize as faithful to the constraints and demands of their respective disciplines. Thus does development ethics serve as a 'means of the means', as a moral beacon illuminating value choices implicit in the instrumental means invoked by decision-makers.

Development ethics assigns relative value allegiances to essential needs, basic power relationships and criteria for determining tolerable levels of human suffering in pursuing social change.

Development ethics as 'means of the means' differs sharply from an ethic of pure efficiency, which is but the rationalization of existing interests. The difference lies between two readings of the dictum 'politics is the art of the possible'. One interpretation is that politics manipulates possibilities within fixed parameters; the second is that politics is the art of creating new possibilities, altering the parameters themselves. Development politics consists essentially of creating new possibilities, not merely in reallocating resources of power, influence and wealth within given boundaries. In practice, this means preferring strategies, programmes, and projects (even modes of reaching decisions) which assign more importance to ethical considerations than to mere technical criteria of efficiency.

In developmental decision-making arenas, three different rationalities converge: technological, political and ethical (Goulet, 1986). Each has a distinct goal and a preferred modus operandi. Poor decisions result when one rationality treats the other two in reductionist fashion, seeking to impose its goals and procedures on the others. If technical rationality holds sway, decisions easily prove to be neither politically feasible nor ethically worthy. Conversely, the triumph of political logic without due regard for the other rationalities may lead to decisions which are technically catastrophic or morally repugnant. Good decisions require many qualities: and not all of these can emerge from a unilateral application of a single rationality.

The essential task of development ethics is to render decisions and actions humane, to ensure that painful changes launched under banners of development do not result in anti-development, which exacts undue sacrifices in personal suffering and societal well-being – all in the name of presumed market efficiency imperative or absolutized ideology of growth as progress.

Culture in development

'Culture' is the living sum of meanings, norms, habits and social artefacts which give one identity as a member of some visible community which has its own way of relating to the environment, of identifying friends, enemies and strangers, and of deciding which values are, or are not, important to it. Essential features of any culture are the definitions it makes of its basic needs and its preferred modes of meeting these. Equally important are the tools which communities employ to relate to their natural and artificial environments: instruments to process materials as well as organizational principles to govern social interactions. Living communities of culture display three essential characteristics.

- A common system of signifying and normative values. Signifying values assign meaning to existence; normative values supply rules as to how life should be lived.
- Some shared basis upon which people identify themselves as members of a group: a common territory, history, ancestors, language, religion or race.
- The will or decision to be self-identified primarily as a member of a given community.

Clearly, not all communities of culture enjoy great economic, social and political vitality: some survive within larger societies merely as marginal adornments thereof or as vestigial relics of the past. Obviously, living communities are not necessarily or always sovereign political units.

How does culture relate to development? What are the cultural dimensions of development and the developmental implications of culture? Cultural diversity has been a precious heritage since the origins of human history. To hundreds of millions of their members, cultures have brought a sense of identity, an ultimate explanation of the meaning of life and death, and assigned them a place and role in the cosmic order of things.

Cultural diversity is assailed by powerful homogenizing forces which relegate many vulnerable cultures to purely ornamental, vestigial or marginal positions in society. The first standardizing force is technology, especially media technology, a potent vector of values such as individualism, consumerism and economic reductionism. A second standardizing force is the modern state, a bureaucratic, centralizing, and legalistic institution inclined to assert control over ideas, resources and 'rules of the game' in all spheres of human activity. A third standardizing force is the spread of the business model of managerial organization as the one best way of making decisions and co-ordinating actions in all institutions. Although these standardizing influences produce massive cultural destruction and dilution, their very pervasiveness gives rise to growing manifestations of cultural affirmation and resistance.

This is why pluralistic development strategies are needed, both domestically and internationally. Economic growth is a legitimate development objective; so are distributional equity, the institutionalization of human rights, the pursuit of ecological health and the fostering of authentic cultural diversity. Consequently, in all societies, development policies, programmes, and projects must negotiate some optimal mix of these diverse (and sometimes conflicting) objectives. No single one should be absolutized or given unconditional primacy over the others.

Authentic development is the construction by a human society of its own history and destiny, its own universe of meanings. Yet most developing societies are forced to work out their destiny in conditions which subject them to multiple destructive influences operating under the banners of modernity, development or progress.

How does any cultural community preserve values essential to its identity and integrity while changing social conditions to improve the quality of life of its people? Every society formulates a strategy for its survival. This strategy, the group's 'existence rationality' (Goulet, 1985), embraces numerous values, some lying at the core of identity, others rippling out into widening circles away from this centre on the outer margins of that core. When organized into a single pattern, these different values form a value system which possesses both unity and meaning.

Rationality is not synonymous with the modern scientific method. Many attitudes and actions which modernists view as irrational, superstitious or uncritical are, when properly understood in their true context, eminently rational (Goulet, 1981). Every society makes sense of reality in accord with the information and interpretative 'filters' available to it, and with its effective access to material and technical resources. Development processes suddenly introduce dramatic new possibilities in both these domains. More important than the rapidity with which modern forces impinge upon traditional cultures, however, are the social structures and contexts within which changes are proposed or imposed. In the last four centuries, western technology has swept the world, thus introducing modernity to Asia, Africa and the American continents. Its mode of domination through military and political conquest, slavery and commercial competition created social instruments which buttress global technological diffusion. Even after slavery and overt colonialism were abolished, many forms of dependence set in place in the age of colonial expansion were preserved by mechanisms of economic commercialism. Consequently, traditional value systems in developing countries must still wage an uphill fight not only to win recognition from others, but also to protect their own fragile self-esteem. Hence the search for alternative development strategies requires the restoration of self-esteem and re-legitimizes indigenous values (Verhelst, 1990).

Development strategies

A consensus is now beginning to form around three essential components of alternative development strategies:

- meeting basic human needs of all as a first priority;
- using resources in ways which enhance long-term sustainability; and
- solving problems by building on local values instead of grafting solutions from outside.

Quite obviously, these normative ideals as to how development is to be pursued run counter to the dominant view of most 'developers' – theorists, practitioners, political leaders, technicians, business personnel, project managers and even some poor populations themselves. All seem firmly committed to development as the pursuit of maximum economic growth – this with little regard for sustainability except in the short term, and with a frankly elitist/expert disdain for local values. Ethical strategies promoting a sound and sane form of development obviously must swim against the mainstream. How then can a change in direction for development be achieved? Clearly, change strategy calls for an informed critique of 'development' as presently conducted, together with sustained extensive public education and mobilization in support of alternative directions (Goulet, 1995). Such efforts face numerous obstacles (Goulet, 1983a) and it is unrealistic to expect an immediate large-scale change in direction. Accordingly, a process of 'creative incrementalism' recommends itself. Gradual innovative problem-solving need not be palliative, however, addressing itself merely to correcting symptoms. If properly designed and well executed, it may contribute to systemic transformation (Goulet, 1970, 1977, 1983b). An alternative development strategy will place heavy reliance on public policies which provide incentives to reward, materially and morally, those economic actors engaging in basic needs/sustainability/participatory development activities (Goulet, 1989). One indicator that normative development ideals presented here may not be unrealizable lies in the growing number of business firms which explicitly commit themselves to operate in an ethically responsible fashion by redefining their mandate as 'optimizing' rather then 'maximizing' profit, so as to assure ecological and social sustainability, and the protection of local values (Schmidheiny, 1992, 1996).

One author in the present volume, an ethicist/practitioner specializing in issues of animal production, offers illuminating testimony in this regard (B.E. Rollin, 1997, personal communication). I asked whether animal producers could adopt ethically sound practices if the change 'would cost them something'. In reply, Rollin cited instances of specific breeders and husbanders who did change to a more 'humane' mode and found it both spiritually rewarding and economically viable. They professed themselves to be 'good enough farmers to make animal raising work under the new system.' Rollin further reported that an entire gathering of animal farmers in Ontario, Canada, raised their hands when asked if they agreed with the animal ethic he had sketched out and if they understood that this ethic constitutes an indictment of the system under which they presently work.

Local values and traditional wisdom must not be romanticized; traditional wisdom may remain grossly deficient. One must ask whether old wisdom, although still influential in developing societies, is capable of articulating a unity of meaning for today's totality. Wisdom, the unity of meaning of totality, is gained after one has faced complexity and multiplicity. In this, it differs from naïveté or oversimplification, which is unity of meaning achieved by avoiding complexity, diversity and radical challenge. In traditional wisdom, formulated in ages more static and less tolerant of pluralistic meaning systems than ours, dogmatism was rampant because accepted meaning systems were not subjected to criticism by competing systems. Moreover, traditional wisdom was framed to guide societies through slow changes over long periods of time. In contrast, the modern world is marked by rapid change and an ever-shrinking time lag between the impingements of change and the deadline faced by societies to react to change. If they are to serve as meaning systems suited to present conditions, ancient wisdom must confront the challenge posed by technological rationality. Reciprocally, if technological rationality is to find an organizing nucleus of meanings around which to make sense of total reality, it needs to dialogue with traditional wisdom. Modern experts often complain that they suffer from insufficient knowledge; in reality, however, they are drowned by too much information, all the while lacking capacious explanatory frameworks. They have no wisdom to match their sciences, no unifying insights to distinguish what is important from what is not. Two varieties of one-eyed giants (Goulet, 1980) abound: those who uncritically accept naïve traditional wisdom and those who promote scientific rationality in a reductionist spirit.

An immense thirst for wisdom exists in developed societies, matched by a no less acute hunger for scientific knowledge in developing countries. Neither mode of knowledge can claim exclusivity: to pursue one to the neglect of the other, or to subsume one form of knowledge under the other, leads to damaging illusions in the realms of knowledge and social policy.

No absolute claims can be made that all cultural groups have the right to survive or to give full institutional expression to their cultural values. Moreover, cultural rights should not be defined in static fashion, but critically re-evaluated in the light of changing development needs. Nevertheless, cultural diversity, within limits yet to be set and agreed to, must be viewed as a community right in the face of the standardizing impact of technology, of its rationality model, of its dominant development prescriptions. Correlatively, if a sound defence of cultural rights is to be mounted, all modes of rationality must be relativized. Technology must be demystified and not idolized as some new Moloch permitted to devour all values standing in its way. The rationality inherent in modern technology is simply one among many possible modes of rationality. More importantly, the non-instrumental treatment of indigenous values must become the dominant approach used by problem-solvers and resource planners. This approach acknowledges that cultural values are essential to people's identity and their sense of meaning, and to their continuity with life around them.

Studies conducted by the British sociologist Peter Marris (1974) suggest that the best way to innovate is to link change to earlier patterns of meaning and significance. Tradition thus becomes an optimal matrix out of which modernity can emerge. However, can the adoption of change strategies overtly seeking continuity with the past suffice to create models of modernity which protect traditional values? A small number of nations are overdeveloped because they waste resources and exploit others. A larger number of nations are underdeveloped: they lack the resources to supply their masses with minimum material well-being. Neither type of society has found a satisfactory institutional answer to the problem of providing essential goods to all in a way that favours human development.

Development means providing all with essential goods needed for living the good life in society so that there can exist that political friendship referred to by Aristotle (in McKeon, 1973) even in the midst of divergent interests and competing ideals. Furthermore, development requires the wise stewardship of resources and a healthy relationship to the environment. By these measures, no national society is fully 'developed.' What justification exists, therefore, for privileging one model of rationality, the highly aggressive model which has characterized western development, over all others?

If, on the contrary, there is room for many models of rationality, there is also room for cultural diversity. This is to say that no solid defence of cultural diversity is possible without an epistemological defence of multiple models of rationality. The cultural rights of societies rest on prior cognitive rights (Goulet, 1987).

LIVESTOCK, POVERTY AND QUALITY OF RURAL LIFE

Ethical choices relating to animal production must be made in several spheres:

- treatment of animals: breeding, raising, confinement and feeding systems, transport, slaughter, marketing;
- health and well-being of animals and of humans;
- environment and pollution from disposal of animal wastes, odour contamination, destruction of species and their habitats;
- quality of rural life and poverty.

On the issue of rural life and poverty, there are two system-wide ethical questions which press for attention. These two questions may be posed in the following forms.

1. Do large-scale livestock operations threaten with extinction the rural way of life as a variety of human culture?
2. Do large-scale operations 'crowd out' alternative uses of land and water better able to meet basic needs of the poor?

To illustrate how ethical decisions can be made in these several domains, I now highlight values and countervalues present in prevailing modes of livestock production and value questions posed in addressing poverty and the quality of rural life.

Livestock

Livestock production is both an important economic activity and a way of life or cultural mode of existence. A few statistics suffice to reveal the importance of animal production in the world economy.

- World meat production in 1996 totalled 195 million tonnes (Mt), up from 192 Mt in 1995.
- The pattern of world meat production is shifting as beef production, which is largely grass-based, presses against the limits of the world's grazing area. In 1950, poultry production was scarcely a third that of beef; rapid growth since then pushed it upward, and in 1996 it overtook beef for the first time in history (Brown, 1997).
- In countries classified as low income economies by the World Bank (1997), agriculture can account for over 50% of gross domestic product (GDP) (Tanzania 58%, Laos 52%, Albania 56%).

- In high income economies, although agriculture's relative share of GDP is low, absolute values are high. In the USA alone, where agriculture accounts for 2% of GDP, over 5 billion animals are slaughtered each year for food (Gruen, 1994).

Large-scale processing of animals

Large-scale processing of animals is conducted in factory chain systems which are ethically reprehensible on several counts: needless cruelty to animals, neglect of worker safety and disregard for the health of final consumers (Johnson, 1991). That these evils are part of 'normal' operating procedures may be seen from the following description by an animal scientist (Skaggs, 1986) of typical slaughtering operations in the USA.

> Iowa Beef's Holcomb, Kansas, plant, America's state-of-the-art facility, sprawls over fourteen acres. Every day cowpunchers push 3,700 head of cattle into a chute that feeds its disassembly line with live raw material. As soon as a steer enters the building, it is automatically zapped by a pneumatic gun that fires a yellow pellet into its skull, stunning the animal, which stumbles to its knees, glassy-eyed. A worker hooks a chain onto a rear hoof, and the comatose beast is mechanically yanked from the platform to hang head down. 'The kill floor looks like a Red Sea', a visiting journalist wrote: 'Warm blood bubbles and coagulates in an ankle-deep pool. The smell sears the nostrils. Men stand in gore with long knives slitting each steer's throat and puncturing the jugular vein. Each night the gooey mess is wiped away from the red brick floors and galvanized steel as required by federal regulations'.
>
> The dead animal, moved steadily by a chain hoist, passes rudimentary disassembly stations consisting of whirling machines and sweating men and women. A skinning machine strips off the hide. Then the carcass is decapitated, the tongue split and removed, all parts being placed on hooks attached to the moving chain. The carcass is gutted, the entrails being inspected and then dropped into stainless-steel containers for eventual use in pet food and other by-products. Disemboweled, the half-ton carcass is pulled through a mechanical washer, quickly examined by an employee of the USDA Food Safety and Inspection Service (FSIS), and split in two by a team of workers manoeuvring motorized saws that rip through bone in seconds. Halves are weighed, washed again, wrapped in sanitary cloth shrouds, and stored overnight in a huge chilled meat locker. (pp. 191–192)

Abusive practices, when occurring in market economies, are rarely corrected through self-imposed change by economic actors with vested interests in continuing those practices. When it occurs, change is usually imposed by regulatory legislation. In political democracies, legislation is itself induced by pressures from a concerned public aroused to the existence of abuses by ethically concerned researchers. Over a decade ago in the UK, such researchers brought to public notice and to the attention of the government, reprehensible general practices in the animal production industry. One influential researcher, Ruth Harrison (1987), testified point-edly on abusive slaughtering practices in the following terms.

> I have seen slaughter in many parts of the world, and I was one of the FAWC [Farm Animal Welfare Council] team which studies it in this country and overseas. I have watched while obviously imperfectly stunned, still conscious animals have been shackled, hoisted and suspended with the whole weight of their bodies pulling on the muscles of one leg, have their throats cut, and continue to hang until they bleed to death. I have also watched whilst deliberately unstunned, fully conscious animals, have their throats cut and are shackled in the same way, their heads and injured throats bumping along the ground until they are hoisted and hang to bleed to death. I have watched, and wondered how we can possibly do this to animals we know to be conscious. What sort of people are we?
>
> Regulations are desperately needed to prevent this sort of cruelty – for cruelty it is. Yet the red meat slaughter report came out over three years ago and no regulations have surfaced yet and some of our recommendations for regulations have been watered down to voluntary codes or rejected altogether. (pp. 16–17)

Harrison contends that animals are entitled to live in conditions where they can at least deploy their natural activities of stretching limbs and wings, rubbing and cleaning without being confined and deprived of light. Although cruel confinement practices are most general in poultry- and veal-raising operations, they are also widespread in beef and pig production. Harrison (1987) considers that humans have a responsibility to apply welfare criteria to farm animals whether they are being reared as replacement animals or for slaughter, whether they are breeding stock or fattening stock, expensive animals or cheap ones. Sometimes, she contends, it becomes necessary to say: 'enough is enough,' and 'either agree to pay more for techniques that do not cause suffering or forego the product altogether' (p. 19).

Imprudent use of animal parts

Ethical questions arise regarding the use of animal parts in ways that jeopardize the health of human consumers. Bovine spongiform encephalopathy (BSE) occurred in cattle in the UK in the mid 1980s. The condition, caused by aberrant prions, has also been called 'mad cow disease' notwithstanding semantic debates among specialists about the accuracy of calling BSE a disease since prions are not pathogens. Cases of BSE in cattle in the UK still continue in 1999, although the incidence is slowly declining. This terrible problem revealed the dangers of using sheep and cattle offal as cattle feed and opened awareness to the growing scientific evidence of trans-species migration of diseases within the animal kingdom. In the UK, it has now been confirmed that BSE, which is a bovine condition, has also made a cross-species migration, this time to humans who ate meat from animals suffering from BSE. The new human condition is known as new variant Creutzfeldt–Jakob Disease (nvCJD). By the end of 1998, 38 people had died from nvCJD in the UK. Since the lead-time between eating affected meat and the appearance of the condition in humans is unknown, accurate predictions of the final human cost are not possible at the time of writing in March 1999. This terrible experience dramatizes the existence of dangers whose ultimate consequences are not known, and perhaps not knowable in the foreseeable future. In this situation, the only ethically responsible course is to assure safety or prevention and avoid the gamble that no harm will result. This approach to biotechnology is known as the 'precautionary principle' (Graham and Wiener, 1995).

Family farms

Family farming, and small-scale animal husbandry on the non-factory farm model, embody a particular cultural way of life whose values are threatened with extinction by the spread of large-scale factory farming of animals. The ethical issues raised here may be summarized conveniently around debates over the desirability of saving the family farm. Admittedly, problems of definition appear. Although the US Department of Agriculture (in its agricultural census) gives an official designation of 'family farm' on the basis of the volume of annual sales, not of ownership or employment structures, the term is generally used as a synonym of a small farm. One advocate, Wendell Berry (1987), defines a family farm as:

> a farm small enough to be farmed by a family, and one that is farmed by a family – perhaps with a small amount of hired help.... By the verb 'farm' I do not mean just the production of marketable crops, but also the responsible maintenance of the health and usability of the place while it is in production. A family farm is one that is properly cared for by its family ... The idea of the family farm, as I have just defined it, is conformable in every way to the idea of good farming: farming that does not destroy either farmland or farm people. The two ideas may in fact be inseparable ... land that is in human use must be lovingly used; it requires intimate knowledge, attention, and care. (pp. 347–349)

Five kinds of arguments, based on ethical or religious grounds, are advanced on behalf of family farms. Appeals are made to values of emotion, efficiency, stewardship, cultural identity, and responsibility. Comstock (1987) takes the argument from responsibility as a synthesis of all, in as much as it:

> draws upon elements of the arguments from emotion, stewardship, and cultural identity....
>
> Because of our duties to future generations, we have a moral obligation to preserve the fertility of soil and the purity of air and water. This requires anticipation, imagination, and vision; it requires looking to the future. But

there are two possible objections to this claim. First, we do not know how to represent the interests of future generations. This philosophical objection should not detain us; just because we do not now know how to represent the interests of unborn people does not mean that we should not try. Second, future generations may develop hydroponic – soil-less – techniques of raising food. This objection rests on a technological possibility that is remote. Even if university scientists were to develop the relevant technologies, the financial burden of constructing that sort of agriculture to feed the world's billions would be prohibitive. For reasons of this sort, I think it safe to say that we have the obligation to look ahead and to do everything we can to preserve clean air, water, and soil.... When it comes to agriculture, the family farm seems to be a reasonable arrangement for meeting these obligations. (pp. 415–416)

Analogous arguments may serve to defend the traditional rural way of life and of animal husbandry as a particular form of culture. Diverse cultures constitute a valuable human patrimony which should not be sacrificed on the altar of maximum economic gain or some putative (albeit flawed) notion of productive efficiency.

Living communities of culture need a minimal economic base which gives them some measure of control over the speed at and direction in which they will develop their resources. In the absence of minimum economic security, a cultural community cannot gain mastery over its destiny: it will be thrust into the role of an eternal supplicant of resources, to be tolerated only on the wider society's terms.

The rural way of life, with its distinctive approach to animal husbandry, is a form of culture worth preserving and activating, because, in addition to its intrinsic value, it has important socio-economic benefits. It promotes employment or, more precisely, livelihoods, and acts powerfully to reduce the mass exodus of rural poor to overburdened cities incapable of providing minimal amenities to all. In short, a development policy which promotes economically dynamic rural living, with farming, animal husbandry and small, technologically appropriate industry, appears to be necessary for achieving economically just (i.e. distributively equitable) development.

Since values of improved quality of life – values of authentic development – at times conflict with traditional values, one must choose which values to endorse. Even when change strategies are designed expressly to minimize value sacrifices, however, there is no way to avoid conflicts. The proper way is to steer a middle course between promoting change without regard to value costs, and romantically imagining that value sacrifices can be totally avoided.

The issue of scale

The central issue is that of economic scale. Economists such as E.F. Schumacher (1973), Fred Hirsch (1976), Ezra Mishan (1993) and Herman Daly (1996) point out that conventional arguments in favour of 'economies of scale' omit calculating the 'diseconomies' attaching to large scale: environmental destruction, elimination of livelihoods and concentration of economic benefits in ways leading to socially intolerable disparities. One major diseconomy attaching to large scale is the danger to public health. Johnson (1991) explains that:

> [A]s beasts from larger geographical areas mingle at auction marts, so opportunities for the rapid spread of infection increase. And as slaughterhouses handle larger numbers of animals, and go over to more production line methods, so the number of carcasses contaminated from a given bacterial source will tend to become larger. In 'line processing' the animals are stunned, bled, eviscerated, skinned and cleaned by different operators as they are moved through the plant mechanically on an overhead conveyor system. Each operator handles many carcasses using the same tools and equipment, and cross-contamination is difficult to control, particularly at high rates of throughput. The conditions in British slaughterhouses have been criticized repeatedly by EC inspectors, who have claimed they show a general lack of awareness of basic hygiene, which allows gross contamination of meat during carcass dressing. Production lines often run too fast, and staff move too freely between 'clean' and 'dirty' areas. (p. 86)

Arguments over the merits of the family farm apply more broadly, in analogous fashion, to farming practices favouring the economic vitality of indigenous groups or cultural communities whose survival is jeopardized by the standardizing impact of a unitary model of economic production. To argue the value of modes of economic activity which preserve such cultural communities is, in effect, to argue the ethical superiority of small-scale communal, more natural and less technologically aggressive modes of animal husbandry.

This line of argumentation admittedly is complex; nor does it lead to simple universal policy prescriptions. A study conducted by Dyson-Hudson (1985) on 'Pastoral production systems and livestock development projects' subsidized by the World Bank in East Africa concludes that some modes of traditional animal husbandry may have no survival value in view of the need to produce larger quantities of food for increasingly urban populations. To this issue of the volume of meat produced must be added another question relating to efficiency. Are large-scale operations necessarily the most efficient? Alan Guebert (1997) a farm and food analyst, cites with approval Michael Duffy, an agricultural economist and associate director of the Leopold Center for Sustainable Agriculture at Iowa State University, who judges that 'farm size and sales-per-farm are poor measures of efficiency and profitability. Our records indicate once a grain farm reaches about 300 acres in size, the cost of production nearly flattens', he says. In other words, the smaller grain farm is as efficient or competitive as the 3000 acre grain farm. The same is true of livestock operations. 'A farm that markets 800 to 1000 head of pigs per year is just as efficient as the mega-operator with 300,000 head', Duffy says.

A theory of consumption

Over 35 years ago, the economist John Kenneth Galbraith (1962) reflected that, '[T]he final requirement of modern development planning is that it have a theory of consumption... – a view of what the production is ultimately for – has been surprisingly little discussed and has been too little missed.... More important, what kind of consumption should be planned?' (p. 43).

Any theory of consumption must make value judgements which differentiate between needs and wants. It must also reach normative conclusions as to the desired composition of the basket of goods and services produced by the economy. To illustrate, the production of animal proteins is high-energy intensive and high-capital intensive (in natural capital, i.e. land and water). What, then, are the true opportunity costs of pre-empting such resources to satisfy the wants (preferences) of consumers who are already well off (meat-eaters) in favour of using less energy and capital-intensive modes of producing vegetable proteins for those in greater need? The 'crowding out' issue centres on protein conversion. One recent analyst, Gardner (1996), contends that:

> [P]erhaps the greatest way to increase food use efficiency is to reduce the world's consumption of meat. The 38% of the world's grain that is fed to animals each year is an inefficient use of cereals. Because a kilogram of feedlot-produced beef represents 7 kilograms of grain, lowering beef consumption could free up grain for direct consumption or as feed to more efficient meat producers (a kilogram of pork represents 4 kilograms of grain, while poultry and fish represent just 2 kilograms). Moving down the food chain would release mountains of grain for consumption by others. If Americans cut their annual grain intake in half, to the level that Italians consume each year, 105 million tons of grain would be saved – enough to feed two thirds of India for a year. (p. 93)

Johnson (1991) argues that:

> [I]f instead of growing crops for direct human consumption, we use the land to grow fodder for herbivorous animals which are then slaughtered for meat, the productivity in terms of human food per acre drops by a factor of ten. And if anyone were rash enough to try farming carnivorous animals for meat – animals which would eat other animals that had themselves fed on plants – yields per acre would be around a hundredth of the usable food that could be obtained from plant staples. If food value is assessed in terms of protein rather than energy, the results are essentially similar, with the protein from an acre of crops ranging from five to fifteen times that from the same acre devoted to meat production, depending on whether cereals, legumes or leafy vegetables are grown. (p. 166)

The larger question here is whether poverty can be reduced simply as a by-product of economic growth by a kind of trickle-down magic, or whether only direct assaults upon poverty by targeting priority allocation of resources to its elimination can succeed. Poverty generally is contrasted with wealth, but critical thought must be given to how wealth is defined.

Defining wealth

In development circles, wealth has come to mean accumulating material or economic goods, personal mass consumption and society's access to an ever-increasing supply of ever-more diverse material goods. Yet genuine human wealth may reside elsewhere, and perhaps economic riches possess only instrumental value in the service of other, qualitative, goods which are constitutive of true human wealth. Testimonies on this quite different conception of wealth appear from diverse sources.

Gandhi
The famous Indian, Gandhi (Das, 1979; Naik, 1983), advocated production by the masses, which brings dignity and livelihood to all, and not mass production, which is production by the few which reduces the masses to being mere consumers of profit-making activities by those few.

Barry Lopez
A student of Native American societies, Barry Lopez (1978), judges 'that a concept of wealth should be founded in physical health and spiritual well-being, not material possessions; that to be "poor" is to be without family, without a tribe – without people who care deeply for you'.

Erich Fromm
The psychologist, Erich Fromm (1976), observes that people always choose one of two modes of living. He makes significant points in the following passages (pp. 15–16).

> [T]he alternative of having versus being does not appeal to common sense. To have, so it would seem, is a normal function of our life: in order to live we must have things. Moreover, we must have things in order to enjoy them. . . .

> Yet the great Masters of Living have made the alternative between having and being a central issue of their respective systems. The Buddha teaches that in order to arrive at the highest stage of human development, we must not crave possessions. Jesus teaches: 'for whosoever will save his life shall lose it; but whosoever will lose his life for my sake, the same shall save it. For what is a man advantaged, if he gain the whole world, and lose himself, or be cast away?' (*Luke* 9: 24–25) Master Eckhart taught that to have nothing and make oneself open and 'empty', not to let one's ego stand in one's way, is the condition for achieving spiritual wealth and strength.

> For many years I had been deeply impressed by this distinction and was seeking its empirical basis in the concrete study of individuals and groups by the psychoanalytic method. What I saw has led me to conclude that this distinction, together with that between love of life and love of the dead, represents the most crucial problem of existence; that empirical anthropological and psychoanalytic data tend to demonstrate that having and being are two fundamental modes of experience, the respective strengths of

which determine the differences between the characters of individuals and various types of social character.

From these and similar testimonies (Goulet, 1996), there emerges a conception of genuine human wealth whose components are:

- the societal provision of essential goods to all;
- modes of production which create 'right livelihoods' for all;
- the use of material goods as a springboard to qualitatively enriching human riches of a spiritual nature;
- the pursuit of material goods in function of their capacities to nurture life and to enhance the being, rather than the having, of people; and
- primacy given to public wealth which, more than personal riches, fosters the common good.

Quality of rural life

Prevailing development strategies are strongly biased against rural living. In the words of Michael Lipton (1976, pp. 13, 16):

> [T]he most important class conflict in the poor countries of the world today is not between labour and capital. Nor is it between foreign and national interests. It is between the rural classes and the urban classes. The rural sector contains most of the poverty, and most of the low-cost sources of potential advance; but the urban sector contains most of the articulateness, organization and power. So the urban classes have been able to 'win' most of the rounds of the struggle with the countryside; but in so doing they have made the development process needlessly slow and unfair. Scarce land, which might grow millet and bean-sprouts for hungry villagers, instead produces a trickle of costly calories from meat and milk, which few except the urban rich (who have ample protein anyway) can afford. Scarce investment, instead of going into water pumps to grow rice, is wasted on urban motorways. Scarce human skills design and administer, not clean village wells and agricultural extension services, but world boxing championships in showpiece stadia. Resource allocations, within the city and the village as well as between them reflect urban priorities rather than equity or efficiency....

> This huge welfare gap is demonstrably inefficient, as well as inequitable. It persists mainly because less than 20 percent of investment for development has gone to the agricultural sector, although over 65 percent of the people of less-developed countries (LDCs), and over 80 percent of the really poor who live on $1 a week each or less, depend for a living on agriculture....

> The misallocation between sectors has created a needless and acute conflict between efficiency and equity. In agriculture the poor farmer with little land is usually efficient in his use of both land and capital, whereas power, construction and industry often do best in big, capital-intensive units; and rural income and power, while far from equal, are less unequal than in the cities.

Ethical reflection conducted in the critical mode suggests the following chain of argumentation:

- a healthy environment is essential to a high quality of rural life;
- a high quality of rural life is impossible without economic vitality in small-scale farming and animal husbandry;
- economic vitality is not possible if rural populations are forcibly expelled to urban centres or reduced to unproductive poverty by the encroachments of large-scale agribusiness and factory farming;
- therefore, supportive infrastructure and development investments should be decentralized – in space and in scale – in order to nurture, strengthen and promote small-scale economically efficient farming practices (including animal husbandry) in rural areas.

A healthy environment is essential to the high quality of rural life because natural surroundings constitute the pervasive material 'envelope', as it were, within which rural people live, move and have their being. More directly and permanently than do their urban counterparts, rural dwellers depend on a healthy and sustainable natural resource base for their livelihood, their health, personal comfort and satisfactions: precisely those elements which make up quality of life. The value of a good environment in rural settings shows forth in salient fashion in a list of basic needs drawn up in 1978 by Sarvodaya, a Sri Lankan self-help development movement. The first of ten essential needs defined by peasant families surveyed is to live in a suitable environment. A 'suitable environment', as understood by Sarvodaya (1978) is:

> [T]he physical, social, emotional and mental environment in which we live. Physical environment includes the house, kitchen, latrine, well. Other water supplies, garden soil, vegetation, the pathways leading to the house, roads in the village, the main roads, air and all other things of a physical character. Included in the social and emotional environment are surroundings impeding spiritual development and concentration and specifically, factors such as noise which contribute to mental disturbances ...
>
> A clean and beautiful environment is one with unpolluted air and soil, suitable for living and devoid of influences that will lead young children going astray and providing physical and mental security and satisfaction.
> (pp. 3–4)

A healthy environment conducive to high quality of rural life is unattainable if economic activities are dominated by large-scale agribusiness and factory farming. Not only are these modes and scales of production highly polluting of air, water and soil; they also contribute, cumulatively and massively, to the destruction of species and habitats, to deforestation and to the ugly standardization of the countryside.

Appropriate scales of animal farming

For these reasons, small-scale modes of production and animal husbandry should occupy a significant proportion of rural sites in order to maximize the protection of nature therein in the pursuit of human livelihoods. These modes of work and productive 'exploitation of nature' are not founded on the principle of maximum, but rather of optimum, productivity and profit-making. The optimal quest for economic gain balances profit-seeking with other values: the protection and revitalization of nature, the care of animals with some sense of kinship with them and some regard for their well-being even if they are ultimately destined for the eating table. Numerous indigenous groups – Quechuas and Aymaras in the Andes, Aborigines in Australia and Fulbe in Senegal – have served centuries-long apprenticeships in treating nature, not as a commodity in the Promethean mode of maximum extraction and exploitation, but in the harmony-seeking mode of stewardship and regeneration.

This is not to advocate a return to unproductive traditional farm practices. On the contrary, pre-modern agricultural societies and farming communities will need to be transformed, in one sense, into post-modern (post-industrial, post mass-scale) units in order to discover social organizational arrangements and technological instrumentalities allowing them to be economically efficient while preserving essential elements of their cultural way of life and collaborative relationship with nature. What is strongly recommended is production diversity in farming – growing both vegetables and grain, combined with running some cattle to provide meat and dairy products, perhaps adding a few pigs and poultry, and even fruit trees and berries. More generally, monoculture, of grains or meat, may be regarded as the curse of modern agriculture and a threat to the future of the agricultural resource base. In contrast, diversified production provides a balanced diet, natural manure (thus reducing the need to resort to chemicals) and a modest production surplus for sale or trade. It is true, as Johnson (1991) observes, that traditional mixed farming is uneconomical.

> ... what is economically favoured in the short term may be quite disastrous over a longer time scale. Unlike those who actually live on and work their home farmlands from generation to generation, the big businesses and banks which dominate the modern agricultural scene have little concern for the long-term sustainability of their projects. If it is profitable now to invest in growing wheat on the fenlands, who cares if tomorrow the fields are a poisoned desert? Capital is mobile, and as the past and present agonies of the American agricultural scene make abundantly clear, money lenders do not hang around when they are no longer making a return. (p. 170)

This line of thinking about development strategy points in the direction of promoting a two-tiered livestock system. One tier is a properly regulated factory-farming sector, needed to supply a significant portion of the large volumes of meat demanded by urban consumers. This requirement holds even on the assumption that meat consumption among economically endowed urbanites may, on grounds of protecting health, plausibly diminish in future years. Argentina, one of the highest per capita meat-consuming nations, has in recent years witnessed a large drop in meat consumption: annual per capita consumption of beef has gone from 89 kg in the 1960s to 70 kg in 1990 and 65 kg in 1992. The following excerpt from the press in Argentina makes a relevant point.

> Nutritionists attribute the decline to the influence of public health educators who have insisted on the need to lower high cholesterol levels. In the past, tradition held that if children didn't eat steaks every day, their physical development would be endangered. Many people now believe there are equally nutritious alternatives.... As eating attitudes have shifted, chicken consumption has risen in the past five years from an average of 10.5 kilos per person to nearly 23 kilos. (*LatinAmerica Press*, 1997; pp. 4–5)

Factory farming needs to be better regulated than at present, if it is to protect the health and safety of animal workers, meet animal welfare standards and ensure environmental soundness in numerous domains – soil and water depletion and pollution, odour contamination, deforestation, habitat destruction and biological species depletion.

The second tier of animal husbandry is that of small-scale and economically efficient family farms and small communal operations (co-operatives, village associations, ejidos, etc.). Unlike large-scale business firms, these economic actors are emotionally and spiritually attached to land. They view nature and animals as valuable in themselves and not solely in instrumental economic terms. These economic actors live in a relatively respectful symbiosis with nature and nature's diversity of life, a biodiversity threatened by the economicist attitude toward treating nature primarily as resource or commodity rather than as value.

The British economist Raphael Kaplinsky (1990) has conducted empirical studies of conditions under which socially, environmentally, culturally, economically and politically appropriate technologies can be adopted in economically efficient ways in manufacturing. These studies shed light on how the second tier of agricultural and animal husbandry may operate. The optimal scale of investment and operations, the mode and level of machinery employed, and the selection of raw materials and intermediate inputs will differ from those prevailing in the dominant mode of agribusiness and factory farming. With proper support from the appropriate combination of what Kaplinsky (1990) calls 'enhancing' state structures and policies, such operations can become economically efficient.

The rural activities portrayed here are simultaneously pre-modern and post-modern. They are pre-modern in preserving social organizations (family farm structures, co-operatives, communitarian production patterns) which pre-date the advent of mass-scale, economically reductionist factory farming and animal husbandry. They are post-modern in going beyond the modern model's narrow conception of economic efficiency, i.e. a profit-maximizing calculus at the level of the firm, to favour a comprehensive view of social and environmental efficiency which treats depletion of natural capital and destruction of human cultural values as costs of production and not as incomes (because in the short term they generate monetary profits for the firm). Over 60 years ago, Lewis Mumford (1934) noted that:

> [R]eal values do not derive from either rarity or crude manpower. It is not rarity that gives the air its power to sustain life, nor is it the human work done that gives milk or bananas their nourishment. In comparison with the effects of chemical action and the sun's rays, the human contribution is a small one. Genuine value lies in the power to sustain or enrich life ... the value lies directly in the life-function; not in its origin, its rarity, or in the work done by human agents. (p. 76)

It lies beyond the scope of the present chapter to specify in detail what precise governmental policies are needed to enable such a second tier of farming and livestock-raising practices to flourish.

Nevertheless, it is clear that these policies bear centrally on decentralizing development investments and on adopting incentive policies favouring this tier (Goulet, 1989). Moreover, experts in agriculture and husbandry need to learn from the traditional wisdom of indigenous farmers who have developed valuable coping mechanisms over long time periods.

CONCLUSION

The biotechnology revolution has introduced a vast panoply of new instruments for enhanced animal production. Advances in biotechnology create novel fast-growing food crops, enhance the output of milk and meat in individual animals, and rapidly increase the size, weight and reproductive capacities of livestock, poultry and marine animals. In many developing countries, however, prevailing social, economic, cultural and physical conditions render unsuitable the adoption of these technological 'improvements'. After evaluating biotech applications in Mexico, a Mexican physiologist and an Iranian agronomist (Hernández-Ledezma and Solyman-Golpashini, 1995), writing jointly, conclude that:

> [A] single formula, applicable to all, does not exist, and even within a specific country various production alternatives are required to meet different social and environmental constraints. As a general rule, production systems should be tailored around the most scarce ingredient, which in most cases is water and land productivity....

> [L]ivestock production depends heavily on soil and water quality – natural resources that are being negatively affected by population increases and industrialization. If the goal is to increase animal productivity in Latin America, then it should be measured in terms of units of animal product per

unit of land instead of units of animal product per animal. Enhanced livestock production should not be achieved by opening more land for livestock utilization, but rather by increasing the efficiency of farms or ranches already in operation – that is, Latin America should adopt a more intensive approach for livestock production rather than the extensive development methods already in existence. However, care must be taken to avoid extracting the maximum benefits from the farms at the expense of dramatically changing fragile ecosystems. (p. 167)

Meat production by factory farming, and its downstream livestock input system, are aimed at maximizing economic growth, not necessarily at generating human development. What is at issue, ultimately, is the kind of development pursued.

Authentic sustainable development

Development strategies now dominant in global arenas aim at maximum growth, the integration of national economies into competitive global markets and the diffusion of technology to every sphere of production. This strategy supplies one implicit answer to the ancient ethical questions about the good life, the foundations of life in society and the stance of humans toward nature. That answer states that:

- the good life resides in the abundance of goods;
- that the basis of life in society is voluntary contracts and pacts to place limits on the harm which competition can produce; and
- that nature possesses value as a resource to be exploited for human purposes.

Moreover, this strategy treats cultural values in purely instrumental fashion: it regards cultural values as mere aids or obstacles to realizing its economic goals. It assumes the universal beneficence of its pursuit of maximum material well-being, of generalized technical efficiency and of modernized institutions.

Nowadays these assumptions increasingly are being challenged and development itself is denounced as an evil thing. The French agronomist René Dumont, writing with a journalist (Dumont and Mottin, 1981), considers the performance of the last 40 years to be a dangerous epidemic of misdevelopment. In Africa, he states, development has simply not occurred. Although Latin America has created great wealth, ranging from sophisticated nuclear and electronic industries to skyscraper cities, this growth has been won at the price of massive pollution, urban congestion and a monumental waste of resources. The majority of the continent's population has not benefitted. According to Dumont (Bergmann, 1987), misdevelopment, or the mismanagement of resources, is the main cause of world hunger and afflicts 'developed' countries as severely as it does developing nations.

Other writers (Sachs, 1992; Buarque, 1993; Escobar, 1995; Wolfe, 1996) strike the same chord: economic growth is often irresponsible, inequitable, destructive and worsens the lot of poor people. Much of what is termed progress is, in truth, 'anti-development' the antithesis of authentic development, which is qualitative improvement in any society's provision of life-sustaining goods, esteem and freedom to all its citizens.

The most absolute attack on development, however, comes from those who totally repudiate it, both as concept and as project. The French economist Serge Latouche (1986, 1993) condemns development as a tool used by advanced western countries to destroy the cultures and the autonomy of nations throughout Africa, Asia and Latin America. The Cultural Survival Movement headquartered at Harvard University likewise has done battle, since its creation in 1972, to prevent 'development' from destroying native cultures. Its founder, the anthropologist David Maybury-Lewis (1987), judges that:

> violence done to indigenous peoples is largely based on prejudices and discrimination that must be exposed and combated. These prejudices are backed up by widely held misconceptions, which presume that traditional societies are inherently obstacles to development or that the recognition of their rights would subvert the nation state. Our research shows that this is untrue. (p. 1)

A clearer view is now emerging as to what constitutes authentic sustainable development (Trzyna, 1995). The World Bank (1992) acknowledges that the 'achievement of sustained and equitable development remains the greatest challenge facing the human race'. Equitable development has not been achieved, however: glaring disparities continue to grow, within and among countries. It will not do, therefore, merely to sustain the kind of development we have presently.

The 1996 World Food Summit, held in Rome, concluded that liberalized trade in food is the solution to the problem of food insecurity (which it defines as lack of access to food). It also urged that more food be produced in the countries and regions where people are hungry, in this concurring with a recommendation made at the UN 1974 World Food Conference, likewise held in Rome.

There may be a contradiction here, for as soon as small local markets in food-deficit countries begin, or threaten, to compete with international corporate-dominated agricultural production and trade, they risk being crushed by the power structure of corporations, banks and collusive governments that dominate the global economy.

The World Commission on Environment and Development (1987) has defined as sustainable, 'development that meets the needs of the present without compromising the ability of future generations to meet their own needs.' This definition, apparently so clear, is none the less fraught with ambiguities. As the economist Paul Streeten notes (1991), it is unclear whether one should: 'be concerned with sustaining the constituents of well-being or its determinants, whether with the means or the ends. Clearly, what ought to matter are the constituents: the health, welfare and prosperity of the people, and not so many tons of minerals, so many trees, or so many animal species' (p. 3).

Although 'sustainable development' has now become the fashionable mantra in international policy circles, it is not self-evident that these two terms are compatible. Sustainability requires simple living in which consumption and resource use are limited. In contrast, development, as conventionally understood, demands continuous economic growth, which may render sustainability impossible by further depleting non-renewable resources and polluting the biosphere.

Development is the major arena where normative, institutional and policy answers are given to the basic ethical questions (the good life, the just society, the stance toward nature) which shape culture. Societies make choices as to what kind of development to pursue as a vision of the good life, and how to pursue it (by what strategies, at what cost). These choices affect the mode of livestock raising they promote, the manner in which they address poverty and the value weighting they attach to environmental health and cultural diversity in building a high quality of rural life.

SUMMARY

Development ethics has two roles. These are to assess competing ends enshrined in diverse strategies and to lay bare values and countervalues in means employed by technical, political and managerial decision-makers in development.

Three functions of culture are described: providing meaning, identity and a place in the universe. Further, cultural diversity, now threatened by standardizing development, is defended as valuable.

Against this background of the operations of ethics and culture in development, ethical choices relating to animal production in numerous spheres are analysed. Factory farming increases poverty and damages rural quality of life. 'Authentic human development,' multidimensional qualitative well-being, is profiled as ethically superior to the economically reductionist form of development now dominant. A normative view of genuine human wealth is presented.

A two-tier model of animal husbandry is presented: well-regulated and limited factory farming; and revitalized small-scale farming characterized by production diversity. This two-tier model is necessary to promote authentic development, foster a high quality of rural life and attack mass poverty at its roots.

ACKNOWLEDGEMENTS

My thanks to Martin M. McLaughlin, Center of Concern (Washington, DC), for helpful suggestions in the preparation of this text.

REFERENCES

Bergmann, B. (1987) René Dumont on misdevelopment in the Third World: a 42 year perspective. *Camel Breeders News* Spring.

Berry, W. (1987) A defence of the family farm. In: Comstock, G. (ed.), *Is There a Moral Obligation to Save the Family Farm?* Iowa State University Press, Ames, Iowa, pp. 347–360.

Brown, L. (1997) *Vital Signs 1997.* W.W. Norton, New York.

Buarque, C. (1993) *The End of Economics? Ethics and the Disorder of Progress.* Zed Books, London.

Comstock, G. (1987) *Is there a Moral Obligation to Save the Family Farm?* Iowa State University Press, Ames, Iowa.

Daly, H. (1996) *Beyond Growth: The Economics of Sustainable Development.* Beacon Press, Boston.

Das, A. (1979) *Foundations of Gandhian Economics.* Center for the Study of Developing Societies, Delhi, India.

Dumont, R. and Mottin, M.-F. (1981) *Le Mal-développement en Amérique Latine.* Les Editions de Seuil, Paris.

Dyson-Hudson, N. (1985) Pastoral production systems and livestock development projects: an East African perspective. In: Cernea, M. (ed.), *Putting People First, Sociological Variables in Rural Development.* Oxford University Press, New York, pp. 157–186.

Escobar, A. (1995) *Encountering Development: the Making and Unmaking of the Third World.* Princeton University Press, Princeton, New Jersey.

Fromm, E. (1976) *To Have or To Be?* Harper & Row, New York.

Galbraith, J.K. (1962) *Economic Development in Perspective.* Harvard University Press, Cambridge, Massachusetts.

Gardner, G. (1996) Preserving agricultural resources. In: Brown, L. *et al.* (eds), *State of the World 1996.* W.W. Norton, New York, pp. 78–94.

Goulet, D. (1970) *Is Gradualism Dead? Ethics and Foreign Policy Series.* Council on Religion and International Affairs, New York.

Goulet, D. (1977) Beyond moralism: ethical strategies in global development. In: McFadden, T.M. (ed.), *Theology Confronts a Changing World.* Twenty-Third Publications, West Mystic, Connecticut, pp. 12–39.

Goulet, D. (1980) Development experts: the one-eyed giants. *World Development* 8, 481–489.

Goulet, D. (1981) In defence of cultural rights: technology, tradition and conflicting models of rationality. *Human Rights Quarterly* 3, 1–18.

Goulet, D. (1983a) Obstacles to world development: an ethical reflection. *World Development* 11, 609–624.

Goulet, D. (1983b) Goals in conflict: corporate success and global justice? In: Williams, O. and Houck, J. (eds), *The Judeo-Christian Vision and the Modern Corporation.* University of Notre Dame Press, Notre Dame, Indiana, pp. 218–247.

Goulet, D. (1985) *The Cruel Choice.* University Press of America, New York.

Goulet, D. (1986) Three rationalities in development decision-making. *World Development* 14, 301–317.

Goulet, D. (1987) Culture and traditional values in development. In: Stratigos, S. and Hughes, P.J. (eds), *The Ethics of Development: the Pacific in the 21st Century.* University of Papua New Guinea Press, Port Moresby, New Guinea, pp. 165–178.

Goulet, D. (1989) *Incentives for Development, the Key to Equity.* New Horizons Press, New York.

Goulet, D. (1995) Developpement mondial: strategies ethiques. In: Dufour, J.L., Klein, Proulx, M.W. and Rada-Donath, A. (eds), *L'Ethique du Developpement: Entre l'Ephemere et le Durable.* Groupe de Recherche et D'Intervention Regionales, Chicoutimi, Canada, pp. 77–89.

Goulet, D. (1996) Authentic development: is it sustainable. In: Pirages, D. (ed.), *Building Sustainable Societies.* M.E. Sharpe, Armonk, New York, pp. 189–206.

Graham, J.D. and Wiener, J.B. (1995) *Risk vs. Risk, Trade-offs in Protecting Health and the Environment.* Harvard University Press, Cambridge, Massachusetts.

Gruen, L. (1994) [From] Animals. In: Gruen L. and Jamieson, D. (eds), *Reflecting on Nature.* Oxford University Press, New York, pp. 281–290.

Guebert, A. (1997) USDA commission needs to ask: What is a farm? In: *The South Bend Tribune.* 30 August, 1997.

Harrison, R. (1987) *Farm Animal Welfare, What, If Any, Progress.* The Hume Memorial Lecture, Royal Society of Medicine, UFAW, London.

Hernández-Ledezma, J.J. and Solyman-Golpashini, V. (1995) Manipulation of gametes and embryos in animal biotechnology's impact on livestock production in Latin America. In: Peritore, N. and Peritore, A.K. (eds), *Biotechnology in Latin America.* Scholarly Resources, Wilmington, Delaware, pp. 147–172.

Hirsch, F. (1976) *Social Limits to Growth.* Harvard University Press, Cambridge, Massachusetts.

Johnson, A. (1991) *Factory Farming.* Basil Blackwell, Ltd, Oxford, UK.

Kaplinsky, R. (1990) *The Economies of Small, Appropriate Technology in a Changing World.* AT International, Washington, DC.

LatinAmerica Press (1997) 29:12, 3 April, 1997.

Latouche, S. (1986) *Faut-il Refuser le Développement?* Presses Universitaires de France, Paris.

Latouche, S. (1993) *In the Wake of the Affluent Society.* Zed Books, London.

Lipton, M. (1976) *Why Poor People Stay Poor.* Harvard University Press, Cambridge, Massachusetts.

Lopez, B. (1978) *The American Indian Mind.* Quest 78.

Marris, P. (1974) *Loss and Change.* Pantheon Books, New York.

Maybury-Lewis, D. (1987) Editorial letter 'Dear Reader'. *Cultural Survival Quarterly* 11, 1–2.

McKeon, R. (ed.) (1973) Aristotle, politics, Book III, Chapter 9, Paragraph 1280. In: *Introduction to Aristotle,* 2nd Edn. University of Chicago Press, Chicago, Illinois.

Mishan, E. (1993) *The Costs of Economic Growth.* Praeger Publishers, Westport, Connecticut.

Mumford, L. (1934) *Technics and Civilization.* Harcourt, Brace and Company, New York.

Naik, J.P. (1983) Gandhi and development theory. *The Review of Politics* 45, 345–365.

Pollock, D.H. (1980) A Latin American strategy to the year 2000: can the past serve as a guide to the future? In: *Norman Patterson School of International Affairs, Conference Proceedings. Latin American Prospects for the 80s: What Kinds of Development?* Vol. 1. Carleton University, Ottawa, Canada, pp. 1–37.

Sachs, W. (ed.) (1992) *The Development Dictionary: a Guide to Knowledge as Power.* Zed Books, London.

Sarvodaya Development Education Institute (1978) *Ten Basic Human Needs and Their Satisfaction.* Sarvodaya Community Education Series, 36, Moratuwa, Sri Lanka.

Schmidheiny, S. (1992) *The Changing Course.* The MIT Press, Cambridge, Massachusetts.

Schmidheiny, S. (1996) *Financing Change: The Financial Community, Eco-efficiency, and Sustainable Development.* MIT Press, Cambridge, Massachusetts.

Schumacher, E.F. (1973) *Small is Beautiful: a Study of Economics as if People Mattered.* Blond & Briggs, London.

Skaggs, J.M. (1986) *Prime Cut: Livestock Raising and Meatpacking in the United States 1607–1983.* Texas A&M University, College Station, Texas.

Streeten, P. (1991) Future generations and socio-economic development – introducing the long-term perspective. Unpublished manuscript dated January 1991. The shorter, published version of this text does not contain the citation given. Des institutions pour un développement durable. *Revue Tiers-Monde* XXXIII (April–June 1992), 455–469.

Trzyna, T.C. (ed.) (1995) *A Sustainable World, Defining and Measuring Sustainable Development.* International Center for the Environment and Public Policy, Sacramento, California.

UNDP (1990) *Human Development Report.* Oxford University Press, New York.

UNDP (1992) *Human Development Report.* Oxford University Press, New York.

UNDP (1994) *Human Development Report.* Oxford University Press, New York.

Verhelst, T.G. (1990) *No Life Without Roots, Culture and Development.* Zed Books, London.

Wolfe, M. (1996) *Elusive Development.* Zed Books, London.

World Commission on Environment and Development (1987) *Our Common Future.* Oxford University Press, New York.

World Bank (1992) *Overview. World Development Report 1992.* Oxford University Press, New York.

World Bank (1997) *Structure of the Economy: Production. World Development Report 1997.* Oxford University Press, New York.

Intensification of Agriculture and Free Trade   8

Dave M. Juday

Hudson Institute's Centre for Global Food Issues, Churchville, Virginia, USA

INTRODUCTION: THE RESTRUCTURING OF LIVESTOCK PRODUCTION AND TRADE

Animal products demand and human population

Business ethics is always a pertinent topic, and the world livestock sector is indeed a major industry with far-reaching implications. However, the ethics of livestock production is a more complex issue than normal commerce, as human consumption of meat has deeply rooted spiritual as well as economic implications. The meat industry market is constrained not only by cultural preference, but also by taboo and religious practice. Many of the major religions of the world have moral codes governing the consumption of flesh; Hinduism, Islam, Judaism and Catholicism, for example, in approximately descending order of strictness. These religions are predominant in some of the most populous and fast growing countries. Thus as commerce, particularly agricultural commerce, frequently bridges the gaps between different cultures and religions, the ethics of the livestock industry are indeed an interesting microcosm of potential conflict. To explore the ethics of livestock production, we must first understand the global trends in demand for meat protein.

Briefly, world demand for animal protein, meat, milk and eggs, is expected nearly to triple over the next 40 years, as animal protein is basically a value-added commodity. This demand is a function first of economic growth and second of population growth. Population growth will have an impact, but in terms of protein demand it is clearly secondary. In the developing world, the rate of population growth has been decreasing for nearly 30 years. Today, the rate of population growth is three-quarters of the way to stabilization, and it continues to slow down. Since 1965, the fertility rate has fallen from 6.5 births per woman (b.p.w.) to 3.2 births b.p.w. Stability is 2.1 b.p.w. In some countries, the rate has dropped even more dramatically. For example, in Indonesia (87% Moslem), the birth rate has dropped from 5.5 to 2.7 b.p.w. and in Colombia (95% Catholic) the birth rate has fallen from 6.5 to 2.6 b.p.w. The World Bank states that the average income of the developing world has risen by about 3.7% per year since 1980. Empirical evidence suggests that rising affluence, urbanization and education has created a desire for smaller families in much of the world and a trend toward high protein or improved diets. Animal protein consumption has increased by about 50%. Protein is an important element in the human diet, as Dr Panya Chotitawan, Chairman of the Thai poultry producer, Saha Farms, has written: '... food is essential, especially protein, which provides both strength and brain structure. Therefore, consuming sufficient protein will generate a healthier body and promote intelligence'.

Birth rates and protein consumption, in an aggregate sense, are variables of affluence. Typically, as economies grow, they become less agrarian. In many underdeveloped rural agrarian economies, large families are an economic necessity. Large families not only provide labour for farm work, but they supply the parents with a retirement income and support. However, in industrialized economies, people often choose to have fewer children. One simple reason is that mortality is usually lower. Another reason is that increased industrial output has the effect of creating jobs and a more complex interdependent economy which provides various economic 'safety nets', such as savings opportunities and more liquid markets for assets like land and homes. These in turn provide retirement security in the developed economy, replacing the security of large families in less developed countries.

Economic growth

In the post-World War II era, there has been tremendous global economic growth, much of it fuelled by the 50-year evolution of the General Agreement on Tariffs and Trade (GATT), now the World Trade Organization (WTO). GATT has made billions of people in hundreds of economies more prosperous by making more extensive trade possible. Over the past five decades, GATT has started a cycle of re-enforcing growth trends among both developed and developing countries that today is beginning to erase those distinctions.

Consider the remarkable growth in Asia over the early and mid 1990s. Korea has become an export powerhouse, experiencing double digit growth for several years. China's economy has quadrupled in size as its exports have increased tenfold. Taiwan's export-led economy has experienced tremendous growth. Japan was the first Asian economy to rebuild itself through exports after World War II, and the effect of Japan's economic growth on change of diet is a model worthy of study. Since 1965, the rice calories of the Japanese diet have been reduced by 37% and red meat calories have increased by 220%. Consumption of dairy products has increased 123%. Japanese now consume about 55 g of animal protein per capita per day.

Currently, there are about 2.8 billion Asians who consume less than 20 g of animal protein per day. If, as the Asian economies grow, these people follow the Japanese model and if current population projections hold true, then by the year 2030, about four billion Asians will consume the current Japanese per capita level of 55 g of animal protein per day. This represents a 390% increase in animal protein consumption. Note that if per capita consumption patterns were to hold steady and population growth were the sole variable, demand would increase by only 57%. Thus the effect of economic growth overshadows the effect of population growth. Indeed, we already see diet change in several countries. For example, consumption of animal protein in Korea was about 6 g in 1960 and today is 25 g, despite tight controls on meat imports that were lifted only in 1997. Indonesia increased its broiler flock from 600 million to 750 million birds in 1995 alone. The consumption of meat in China has increased more than 10% in each of the past 5 years.

Even in India, where many do not eat any meat, there has been a huge growth in animal protein demand. Since 1980, milk consumption has more than doubled to 65 million tonnes (Mt) per year, and milk consumption continues to rise 4–5% annually. About half of this is consumed as liquid milk, and the balance in other forms. Consumption of eggs reached 28 billion in 1995, which was double that of previous years, and official estimates predict egg production will reach 40 billion by 2000. For religious reasons, many people in India do not eat poultry meat, but do consume chicken eggs because eggs are unfertilized. Likewise, Hindus do not eat beef but drink cow's milk. These practices are, of course, legitimate religious practices, as are halal and kosher slaughter rituals. However, an important point here is the fact that, regardless of the slaughter ritual or whether the bovine animal is a dairy cow or a beef steer or, for that matter a draught animal, the essential impact of animals upon the environment is the same, because livestock need feed which takes land for production.

Therefore, as Asia develops in both population size and in affluence, land will become a more scarce commodity. Asia already has by far the lowest ratio of arable land to human population of any region of the world. Currently, Asia has six times as many people per acre of cropland as the western hemisphere. By 2030, this figure will increase to eight times as many people per crop acre as North America. By the middle of the 21st century, 50% of the world population will live in Asia, with only 31% of world arable land and 23% of world pasture land. Already market land values show the relative availability and productivity of crop land in various countries. According to Knight-Frank Property Management of the UK, 1996 crop land value per acre in Japan was US$101,174; in western Europe it was between US$4000 and US$9500; in the USA, it was US$2500, and in Canada it was US$1200. In Argentina, crop land in 1996 was US$850.

Argentina, one of the world's most promising agricultural nations with the lowest crop land value, makes an interesting case study. The famed Argentine pampas have become crop land rather than pasture land. Currently, 80% of the pampas now support corn, soybeans and wheat. This is a response to the new world demand for feed grains, as well as to the suspension of Argentina's export tax. This switch of land use, in which crops replace cattle on prime land and cattle move to marginal land, is also resulting in the restructuring of cattle production systems with less grazing and more supplementary feeding. In 1998, according to the US Embassy in Buenos Aires, there were about 100 feedlots in operation, and slaughter houses were forming alliances with feedlots to target the export market.

Indeed, we are seeing much of the same in the USA. Pork production has moved toward concentrated feeding operations and, in the process, has largely moved away from the traditional grain belt toward other areas that supply lower wages, cheaper land and proximity to ports. The state of Iowa, in the centre of the USA, was the leading US pig-producing state, primarily due to the production of feed corn. From the spring of 1997, were it not for a 2-year moratorium enacted on new pig farms, the east coast state of North Carolina would have surpassed Iowa in pig production. Among the reasons for this is that North Carolina has lower wage rates, proximity to larger urban centres and to coastal ports and an otherwise shrinking agricultural commodity base (peanuts and tobacco) leading farmers to seek other opportunities. Similarly, pig production has moved westward to arid and semi-arid regions that do not support crops and which are closer to Pacific Ocean ports and the Asian market.

Livestock intensification

The restructuring, or intensification, of the livestock feeding industry, both in the USA and in Asia, is following a pattern first developed in The Netherlands where crop land is scarce and pig operations are located near port cities for ease of importing feed. This model is likely to be adopted in Asian feed operations, for example in countries such as The Philippines for pork production and in Indonesia and Thailand for poultry production. With pork production near the seaports, the Dutch, who traditionally have been one of the ten largest pork exporters in the world, were able to maintain an apparently unlikely position in the export market by exploiting the comparative advantage of their seaport system. Other nations may have comparative advantages in capacity for feed grain production or the size of

their consumer market. Each comparative advantage plays a role in determining the place of a nation in international commerce.

Of course, this restructuring is the definite trend in livestock production
and agriculture generally – intensification and trade, which is the assigned
title of this chapter.

We have now identified significant restructuring trends in the livestock
sector and indeed in agriculture generally leading to intensification and increased trade. The next step is to consider these changes and trends within
an ethical framework. This chapter identifies three ethical issues: food
security, global environmental protection and cultural diversity.

FOOD SECURITY

Principle of comparative advantage

Comparative advantage is the theory of economic allocation that was developed by economist David Ricardo (1965) in the 1800s. Using the famous
case of wine and wheat trade between Portugal and England, Ricardo
argued that countries should specialize in (or intensify) the production of
goods or commodities which they are relatively more efficient at producing
and then trade with the rest of the world to obtain other needed goods. In
practice, many nations traditionally have avoided agricultural trade for
reasons of 'national food security'. In 1947, after World War II when the
original GATT negotiations were taking place, European governments,
which already had a long tradition of crop price supports and import
barriers, were trying to rebuild their war-shattered agricultural economies.
They naturally placed an emphasis on national food production. Meanwhile, in the USA, the government had promised farmers high price
supports after World War II if they accepted food price controls during the
war. During the time period of the trade negotiations, both Europe and the
USA had much larger farm populations than today, making agricultural
trade too tough politically to open food to broad global private-sector trade.
Those moves to ensure food security seemed rational from a national
emotional perspective in the context of recovering from World War II.
Today, on the verge of the 21st century, economic and ecological evidence
mandates that food and agricultural policies must be built on a system of
broader private trade; indeed, perhaps in no other sector of the economy
are the benefits of comparative advantage in trade so pronounced as in
agriculture.

Under the system of comparative advantage, total world production
would be physically greater under a system of trade rather than if each
nation tries to balance its domestic demand and production. As an example,
let us consider two countries, called Country A and Country B. Country A
is wetter and warmer and suited to rice production. Country B is cooler and

drier and suited to wheat production. To keep the example simple, let us assume that each country produces 100 bushels per acre of their specialized commodity, and 50 bushels per acre of the other commodity. Let us also assume that each country has 90 acres in production. If each country tries to be self-sufficient and grow equal amounts of rice and wheat, Country A would grow 30 acres of rice at 100 bushels per acre to yield 30,000 bushels, and it would grow 60 acres of wheat at 50 bushels per acre for 30,000 bushels. Country B would do the reverse, growing 30 acres of wheat at 100 bushels per acre to yield 30,000 bushels, and growing 60 acres of rice 50 bushels per acre for 30,000 bushels. Each country would then have 60,000 bushels of grain produced domestically, making a total production for the two countries of 120,000 bushels of grains.

We now consider what would happen if each country specialized in the crop better suited to their climate and soil type, and then entered a trade agreement. Country A would devote all 90 acres to rice, thus producing 90,000 bushels of rice. Country B would devote all 90 of its acres to wheat and thus would produce 90,000 bushels of wheat. The total production of the two countries would be 180,000 bushels. Under a system of trade, each country could then have 90,000 bushels of grain, including both rice and wheat. This model yields a 50% increase in supplies compared with the scenario of self-sufficiency.

Obviously, the concept of comparative advantage is based on the marginal cost of production, i.e. the cost to produce one more unit of a good or commodity. For example, as the marginal cost of producing more grain increases, a nation gains an advantage by importing the extra grain. It would benefit China, for example, which is already double cropping its Northern Plain, to import supplies to meet its marginal demand for corn. However, as recently as late 1995, the Chinese Minister of Agriculture was advocating a policy of food self-sufficiency. To achieve ongoing self-sufficiency, which may not even be possible, China would have to invest huge amounts of money in dams, land clearing and greenhouses to keep production at its maximum. This type of development and investment would divert financial resources from other more lucrative public and private capital investments, thus raising, in one sense, ethical questions about quality of life.

Even within China's agricultural sector, increasing production of food and feed grains would have significant opportunity costs. For example, textile production is a major employer in China and also a major source of export income that would be forgone if cotton production were sacrificed for corn or other commodities. Moreover, such expensive self-sufficiency insulates the Chinese market from more competitively priced grains produced elsewhere, thus further increasing the costs of feed grains at a time when China is experiencing a tremendous increased demand for meat. Annual meat consumption in China has increased by 20 Mt since 1993. To keep up with this rate of demand, experts predict that China will have to increase the capacity of the manufactured feed industry from 30 Mt year^{-1} to at least 100 Mt, if not 120 Mt, by the year 2000.

A parallel case is offered by US coffee consumption. The USA consumes about 30–40% of world coffee production and imports virtually all its supply. Though the US state of Hawaii does produce some very high quality, prized coffee, it remains to America's comparative advantage to import coffee from other nations such as Indonesia, Brazil, Mexico and Kenya instead of trying to devote all of the state of Hawaii to achieving coffee self-sufficiency for the USA. Like corn production in China, coffee self-sufficiency in the USA could be a goal, but economically it is not worth trying because other places, such as Kenya, can produce marginal units of coffee more cheaply. Even though Kenya does not import much US agriculture production, or other goods in return, it is still to America's advantage, and also to Korea's advantage, for the USA to import coffee from Kenya. The reason is that expansion of coffee production in Hawaii would mean diverting resources which are better invested in the more important tourist economy and hotel services, and this would result in fewer foreign travellers coming to Hawaii. Consequently, the USA would drink more expensive domestic coffee and earn less income from tourism. With less US income, there would be less purchasing power to import Korean cars. Thus, odd as it may seem, under the principle of comparative advantage, there is a certain quantity of imported Kenyan coffee in the USA that equals one Korean-made Hyundai car and, in human terms, all of those goods and commodities represent employment and the basic right to participate in the economic life of society.

The issue of farm size

One of the chief criticisms of broad-scale economic efficiency of the type achieved under the dynamics of comparative advantage is the potential effect on the individual with respect to employment, which in the case of agriculture affects the individual farmer and livestock producer. The case is often made that efficiency drives out the small and vulnerable producer. While the economic plight of the individual is a legitimate ethical question, arguments made against efficiency for the sake of the individual, while emotionally powerful, often do not hold up well to scrutiny. The reason is that the rights of others in society are often set aside under such scenario constructs.

An example of this issue comes from the USA, where the notion that farmers were entitled to a minimum guaranteed income was embodied in a programme of supply control for more than six decades. This programme restricted food production. According to a study by the University of Minnesota, the programme led to a 30% decline in the non-farm rural population. The commodities which were not produced under supply control measures were not stored, not transported and not processed. Consequently, in this case of economic inefficiency, certain non-farmers in rural communities were denied the rights to choose their jobs and sources of income. Likewise, a 1996 law in the USA allowed six states in the northeast of the country to establish a compact to protect the income of regional dairy farmers. The result was an increase in the retail cost of milk in that region, and a reduction in milk consumption among lower income consumers. Given that milk is important in the diet of children, regional dairy farmers were being subsidized at the potential cost of the health of poor children. This domestic US regional cartel, known as the Northeast Interstate Dairy Compact, is in fact a microcosm of the trade policies of various nations which are designed to protect their farmers. India, for instance, limits dairy imports to protect their farmers; yet Indian consumers pay above world average milk prices and, according to 1998 statistics, consume, per capita, only about two-thirds of the recommended minima given by the World Health Organization. Thus, ultimately, the ethics of protecting dairy farmers in India must be weighed against the interests of consumers there, including growing children and pregnant mothers. It must also be weighed against the interests of peasant farmers in Poland, a country with suitable pasture land to have a comparative advantage in global dairy production were global markets more open.

To look at it from another perspective, efforts to guarantee a certain standard of living for producers, such as market cartels or tax subsidies, mean that participating producers are given a special right to claim a certain amount of the financial resources of society. It could be argued that the right ought to be accompanied by an ethical responsibility to tailor their production to the needs of society. The point could be extended to the commodity production of a peasant farmer in the developing world by expecting him to accept some ethical responsibility. For example, let us again consider India and cattle, where the dairy farmers are protected. Would a livestock farmer in India have an ethical right to demand income from Hindu society even if he raises his cattle for beef rather than dairy? Or would he have a prior ethical responsibility to the overall value of a society that considers eating beef an anathema? If it can be demonstrated that there is a responsibility to the overall society, then it also can be argued that applying the dynamics of comparative advantage is indeed a suitable, if not the most suitable, means to meet and uphold that responsibility.

Indeed, economic allocation is often the central organizing principle of many of the great, and not so great, world philosophies. As food is essential to human life, food must be at the centre of the ethical debate over economic allocation. It is too often forgotten in the debate, however, that the means of producing food are not evenly distributed. In fact, crop and pasture land, as well as degree days and water, are poorly distributed between cultures, societies and nations, and, further, are not matched to their expected consumer demands in the coming decades. India cannot produce the milk it demands; New Zealand produces more than it can consume. Thus, trade is a means by which mankind may redistribute more equitably, and in certain cases more abundantly, what nature has not distributed evenly or according to need. Given the validity of these facts, there is clearly an ethical component within the design of trade policies. This means that governments, such as India, the USA, the European Union (EU) and other nations, have an ethical responsibility to ensure that their trade policies are not influenced by the interests of individual sectors within their domestic societies who could seek benefits which would be against the interests of global society as a whole.

ENVIRONMENTAL PROTECTION, BIOTECHNOLOGY AND FEEDING THE WORLD

Food production and natural resources

Food production depends upon natural resources and thus in agriculture we collaborate with the forces of nature. Therefore, we must consider the impact of agriculture on the environment. It is especially important in the case of animal agriculture which is more resource costly. Consider again the example of a dramatic increase in coffee production in Hawaii. Such an increase would not only mean increasing food expenditures as demonstrated previously, but would also result in the destruction of a unique ecosystem and beautiful natural area in the USA. Therefore, this example also raises the need to care for the environment within the issue of expanding agriculture trade.

Ricardo himself used such an example in explaining the benefits of trade. Ricardo noted that when corn production increases, land of inferior quality is used and thus the marginal cost of producing corn increases. Ricardo's example is prophetic in the case of Asia on the eve of the 21st century. Increased agricultural production in Asia will come at great cost, not only financially and economically but also ecologically. Inferior quality land is more economically costly because it is more ecologically fragile and thus less productive. In Asia, where demand for food is growing so fast, most of the land which remains unused is tropical and thus inherently suffers from high soil temperatures and low organic content. In this situation, there are two issues of environmental importance. First, tropical land is home to great biodiversity. As ecologist Michael Huston (1994) points out in his book, *Biological Diversity*, in rain forests and swamps, the tough conditions force wild and plant life into narrow niches, thus producing lots of species. Tropical forests harbour 60—80% of all plant species. Second, tropical land has high erosion potential. The overriding conclusion is that it makes economic sense to preserve the environment through trade and it makes ecological sense to maximize the economy through trade. Lands with the most biodiversity are also the poorest crop lands. By contrast, the best crop lands are home to thriving populations of only a few species. In the USA, for example, more than 100,000 acres were cleared for farming in the corn belt of Ohio and Indiana and not one unique endemic plant species was lost, such plants being important indicators of biodiversity. This example contrasts dramatically with our earlier hypothetical example of introducing large-scale coffee production in Hawaii since clearing tropical forests in Hawaii for coffee production would result in the loss of more than 880 unique indigenous plant species.

Farm costs and ecological costs

Unfortunately, there is a real life example of this marginal production with high ecological costs. Indonesia increased soybean production to feed its expanding broiler flock. The extra production is coming from tropical forest land which is being cleared. Meanwhile, the year 1997–1998 saw a record world soybean harvest, owing to South America's outstanding crop and increased production in the USA. Unlike Indonesian production, the North and South American soybeans will be grown on good crop land with higher yields, lower economic costs and less ecological destruction. The comparative advantage for Indonesia, both in terms of its economy and ecology, shows that it would be advantageous to import soybeans from the Americas where the marginal costs of production are lower.

It is worth re-emphasizing that marginal cost is the important concept underlying the theory of comparative advantage. Thus, agricultural trade based on comparative advantage would not result in Asia losing all its farms and being left with no ability to produce food. It simply means that Asian countries could obtain the additional production to meet increasing demand for food, the marginal increases in demand, in ways that relieve them of great economic and ecological costs. In fact, Asia already has developed the most intensive farming systems in the world. Wet rice culture was the first truly intensive farming system in the world and it has enabled Asian nations to support very dense populations for centuries. However, most of Asian agriculture is now at the point of diminishing marginal returns and, economically, these countries would be better off with expanded trade.

The Philippines example

The Philippines instituted an agricultural policy in 1996 which provides a good example of how trade can and will benefit fast-growing Asian economies, including the domestic farm sector while protecting ecologically fragile wild lands. The previous food policy in The Philippines was aimed primarily at producing enough rice domestically to meet the country's own needs. However, when per capita incomes rose to new levels, the demand for milk, meat and eggs surged as well, just as it had in Japan decades earlier. The result was a rise in corn prices in The Philippines to twice the level on the world market (yellow corn imports were banned to protect white corn production). Despite 3 years of record domestic grain production, demand outpaced supply and, in the second half of 1995, stocks began to run out and prices soared by 30–50% – and the Minister of Agriculture lost his job.

The new Minister of Agriculture, Salvador Escudero, instituted a plan that bears his name and that opens up The Philippines to food imports. Eventually it will levy only one low tariff for all agricultural products. Under the plan, The Philippines expects that, on average, about 5% of its rice will be imported each year; and when rice crops are large, Filipino farmers may export rice. This policy allows farmers to 'stockpile' income and to invest or to save it wisely, instead of perpetuating a system where the government stockpiles rice and pays the farmers a minimum income guarantee.

Under the Escudero plan, change came to the Philippine livestock industry almost immediately. Previously, because of the ban on the import of yellow feed corn designed to protect production of white corn of food grade, the livestock industry had relied on wheat which was more costly and less efficient as animal feed. As a consequence, imports of yellow corn began to soar. Annual imports used to total less than 100,000 t, but in 1997 the import increased to 800,000 t. These imports, of course, help expand domestic livestock farmers' production.

The new Philippine plan deliberately puts farmers under competitive pressure to stimulate better jobs and a stronger comparative advantage for the nation's farmers, even as food prices are lowered for consumers. The increased efficiency in the livestock sector will create rural jobs and stimulate a competitive economic climate for agricultural investment. Comparative advantages, in the long run, will benefit farmers more than government programmes and import protections.

A further example comes from Malaysia. As trade liberalization progresses, imports of livestock, meat, dairy products, corn and wheat have all increased. At the same time, however, Malaysian farmers have enjoyed increased exports of tobacco, cocoa and palm oil, and more than 4% general growth within the agricultural sector.

The Philippines case also illustrates an important political lesson for governments. Food consumers will not support farm policies that subsidize domestic farmers through higher food prices and food supply volatility. Moreover, though farmers may lobby for import protection, maintaining minimum levels of income through price supports and import protections will not keep farmers politically content for long. Many farmers now realize that being able to supply a guaranteed domestic market does not necessarily mean enjoying profits.

In any industry, isolation from competitive pressures typically shows a number of characteristics, including poor cost control, poor marketing and poor use of technology, all of which eat away at profits, both in the present and in the future. Filipino livestock producers now realize that wheat is an expensive animal feed compared with corn. However, under the previous system, they had little control or options. Meanwhile, the high cost of production limited the market demand for meat.

Biotechnology

Technology also plays a role in more effective and efficient distribution of food. In fact, telecommunications, telephones, fax machines and now e-mail, may be one of the more important technological developments for agriculture. Importing countries are in constant contact with producers before and after they plant. This allows more efficient and secure purchasing decisions. However, in addition to support in merchandising, technology is also dramatically changing the way in which agricultural commodities are produced. Already the USA has exported to Europe genetically modified (GM) soybeans, known as 'Round Up Ready' soybeans, that are tolerant of herbicides, and 'Bt' corn that is resistant to certain pests. Unfortunately, the EU, spurred by environmental activists in western Europe, is resisting these products which, if accepted, could further increase the benefits of comparative advantage

Environmental activists in Switzerland including the groups Greenpeace and World Wildlife Fund, spent 6 years in a campaign against biotechnology. That campaign resulted in a referendum, the 'Gene Protection Initiative', that would have forbidden the release of any GM organism into the environment. That prohibition, in effect, would have stopped biotechnology development in Switzerland which is, ironically, a global centre for the industry. The campaign was confronted by scientists and biotechnology companies who argued the economic benefits of its biotechnology industry to Switzerland and promoted in detail the benefits from medical breakthroughs offered by biotechnology. The 'Gene Protection Initiative' ultimately was defeated, but its evolution is worth study.

Indeed, biotechnology did not fare so well 2 years earlier when nearly one in seven Swiss citizens signed petitions circulated by environmental activists opposing the importation of GM foods. This action sent shock waves through the feed grain and oil seed markets. Although Switzerland is not a member of the EU, it is a microcosm of the politics of opposition to biotechnology in western Europe. According to the research of Dr Lynn Frewer of the Institute of Food Research in the UK, the typical responses to poll questions in EU countries to questions such as 'should genetically modified food be developed?' or 'would you consume genetically modified food?' are between 60 and 80% negative. Dr Frewer notes that most respondents are unaware of the benefits of biotechnology and are not familiar with food production. In this case, these polling respondents are not unlike the Swiss electorate before public education was offered during the debate over the 'Gene Protection Initiative'.

Labelling of GM food

The EU now requires labelling of GM food under EU Regulation 1139/98, a regulation that has several implications. Dr Frewer argues, from a psychological perspective, that the labelling is a positive development, as she explained in a symposium in the USA: labelling 'gives consumers choice about something they do not fully understand, putting them in control of their fears, and conveying to the public a message of openness and honesty'. However, the opposite case can be made: labelling of biotechnology is far from open or honest, and it does little more than stir fears. This in itself is an issue of ethics.

Government mandated labels have an imprimatur of warning, for example, warnings about use of alcohol, tobacco and prescription drugs. Thus, mandated labels for GM food communicate an imputed warning. In the absence of a clear and compelling case for public safety, the labels should be considered no more than a trade barrier.

bST

Consider the case of one of the first biotechnology products in the USA, bovine somatotropin (bST). Efforts by the environmental lobby to require labelling, unsuccessful to date, failed because bST is harmless to humans and cannot be proven otherwise. Humans have been safely exposed to natural bST in cows' milk from the time dairy cows were domesticated until pasteurization was perfected. In today's era, pasteurization of milk eliminates any trace of it but, even if humans consume it, bST is a protein and is digested in the same way as eating a piece of meat. Thus, the proposed labelling serves no public health purpose. Therefore, because EU Regulation 1139/98 mandates labelling, the burden is on the EU to communicate the public health purpose of that regulation. Short of that, the Regulation is little more than a trade barrier, which it can be argued is a detriment to environmental protection.

Again, consider the case of bST. It is a fact, well documented in the USA, that the use of bST enables a farmer to produce more milk from fewer cows. Therefore, in addition to cutting costs, and so increasing his profits, a dairy farmer produces less solid waste and uses overall less feed, while producing more milk. Thus with bST, consumers, producers and ecology benefit.

Biotechnology will also contribute greatly in the area of meat production. Pork is the world's most heavily consumed meat, and pork somatotropin (pST) has even greater potential than bST. In fact, pST can produce pigs with 60% less fat and 15% more lean. It took the swine industry more than 30 years to develop a pig that today is 50% leaner than the average pig in 1960. However, with pST, these gains will be achieved in short periods of time, using one-third less feed grain. In 1995, Japan imported 830,000 t of pork. Since pork has a feed-to-meat ratio of about 5:1, the imported pork took about 4.15 Mt of feed. If pST were used, more than 1.38 Mt of feed could have been saved, that is equivalent of about 420,000 acres of average yield US corn land. The amount of land used would be even higher if the feed were grown in most parts of Asia.

Similarly, research and technology continue to increase crop yields. In the USA, the average corn yield in 1930 was 22 bushels per acre; today it is 130 bushels per acre. Eventually it will be 400 bushels per acre according to researchers at Purdue University.

Bioethics and biotechnology

The debate over bioethics is by no means settled, as more detailed analyses explain in other chapters of this book. One of the authors, David Cook has written, 'Just because we are able to do something technically, does not imply that we must do it'. I would like to supplement his point with this statement 'Just because we have not done something previously, does not imply that we must not ever do it'.

In our earlier case of comparative advantage in the allocation of national resources, we defined an ethical issue in the rights of individuals versus society. In this section of the chapter, the ethical issue is a relative balancing of the rights of various people or cultures and our role as stewards or animal husbandmen. As humans together in the world, we share at least one common experience – we must eat to live. This human need, when considered globally from the overall perspective, leads to an ethical responsibility to consider the benefits which biotechnology can confer upon food production. Benefits already apparent include giving livestock the ability to convert feed more efficiently and giving feed crops the ability to produce higher yields. Together, these benefits offer the chance to satisfy the legitimate growing demand for meat, dairy products and eggs with less negative environmental impact.

The production of more food from less acres will enable a world of 8.5 billion affluent people in the future to enjoy the diets they demand without sacrificing wild lands to the plough. Each generation has an ethical obligation to the next generation in terms of environmental stewardship, free trade and intensification. Biotechnology appears to offer a suitable means of fulfilling that obligation.

CULTURAL DIVERSITY

Human diversity

It is true, however, that there is no guarantee that high-yield agriculture offers zero risk of environmental impact, which is the stated reason why the modern environmental movement has eschewed it. The opponents of high-yield, intensive food production offer the alternative of 'sustainable agriculture', a term used to describe a number of traditional and low-input practices. However, the ecological risks of high-yield agriculture are low and getting lower with each new scientific breakthrough. The same cannot be said for most agricultural practices termed sustainable. For example, the International Forestry Centre recently warned that the world may lose half of its tropical forests due to the expansion of traditional farming practices. These traditional practices rely on ploughing farm land which is subject to vastly more soil erosion than the higher yield farming systems supported by chemical fertilizers and pesticides under what has been termed 'conservation tillage'. More importantly, the yields from traditional practice systems are lower and thus need more land to produce the same amount of food. Thus, in a formulation not uncommon to the environmental movement, another author in this book, Gary Comstock, asks rhetorically, 'How much land can 10 billion meat eaters spare for wildlife?' Given the projections of the International Forestry Centre, the answer is in part a variable of farming and production practices. However, even prescriptions for the manipulation of food and livestock production raise ethical conflicts. The Humane Society, in a not uncommon proposition for the environmental movement, stated in testimony to the US Department of Agriculture that 'the fear of (human) hunger should not be placed over the concerns about the environment, ... and the humane treatment of farm animals'.

The population control issue

In this debate, the alternative offered most frequently to the manipulation of food production and distribution practices is regulation of human reproduction rates to serve a status quo of consumption demand. For example, in a speech to the United Nations, US President Bill Clinton said 'to ensure a healthier and more abundant world, we simply must slow the world's growth in population'. Actress–activist Jane Fonda, in a speech to the United Nations Population Fund, stated 'the controversy around contraception and abortion made it politically easier to speak and organize around air pollution, deforestation, toxic waste, and biodiversity while ignoring the role our own burgeoning species places in all of this'. Therefore, following the lead of Clinton and Fonda, top environmental groups successfully lobbied the United Nations during the Cairo population conference in

1994 for a $17 billion pledge for population control. Moves in this direction have caused Nigerian scholars to write of their fear of an 'imminent threat of a Western birth control dictatorship in the Southern hemisphere'. Indeed, this is an ethical conflict among cultures.

Concurrently, major western environmental groups have made 'cultural diversity' a secondary plank on their platform. For example, the Humane Society of the United States has stated in testimony to the US Department of Agriculture, the 'preservation of cultural diversity has inherent value just as does the preservation and enhancement of natural biodiversity'. However, these two priorities, cultural diversity and population control, produce some predictable ethical conflicts. There is a rather significant number of cultures in the developing world, current targets of western population control programmes, whose moral teachings reject artificial contraception. Authentic Islam, genuine Catholicism and the teachings of Hindu nationalist leader Mahatma Gandhi are examples of three diverse cultural forces which share a common, and unequivocal, opposition to artificial means of birth control.

As stated in the opening of this chapter, the rate of world population growth has been decreasing for nearly 30 years and today is three-quarters of the way to stabilization. In short, one could argue mathematically that the fundamental premise of forced global population control is groundless in addition to being ethically questionable. While absolute world population totals are increasing, and will continue to do so for the next 30–40 years, the rate of population growth is now 25% lower than it was two decades ago and, more significantly, it continues a steady decline. Therefore, expanding use of contraceptives would probably make only a marginal difference in the world's birth rate; yet contraception remains a fundamental cultural difference in the way in which different peoples view humanity. Agricultural intensification and free trade offer the best hopes of sustaining wildlife, and a growing population of affluent meat-eaters, without the conflicts of forced population control.

CONCLUSION

Economists call the modelling of economic effect, with all its implications, general equilibrium analysis. It is clear that intensification and free trade are economic trends within the agricultural market. However, as the French economist Leon Walrus, a pioneer in general equilibrium analysis, argued, no market can be understood in isolation. As stated at the beginning of this chapter, the ethical dimensions of food consumption, especially of meat, add additional layers on to the economic analysis. These ethical layers do not offer the same lucidity to all viewers because, just as markets differ, so also codes of ethics are different according to varying cultures and values.

However, because of the physical requirement for food for all humans to live, there is relatively little variance in the elasticity of demand for food from economy to economy, or from culture to culture. In short, food is a necessity and consumers purchase it first. Money, time, intellectual capital or labour that is not spent on food is available for allocation to other goods and services or non-economic pursuits. By way of comparison, Americans spend about 10% of their income on food; the Japanese about 25% and the Chinese about 70%.

Logically, intensification and trade in agriculture are meant to lower costs and expand distribution, which will have the effect of bringing these divergent spending patterns closer and, therefore, of releasing certain other resources into new uses. Obviously, such trade develops interdependence between nations in a general equilibrium model. If China, for example, were to spend only 50 and not 70% of its income on food by buying imported rice from Thailand, wheat from Australia and corn from Argentina, then a new segment of income (14% of total Chinese income) could be spent on other developments. In an ethical sense, this interdependence can, and often does, strengthen the universality of humankind, because interdependent economies are less likely to be adversarial. Such interdependence also maximizes the beneficial purpose of the goods of nature.

While global forces are complicated and nuanced, a cogent case can be made for intensification of agriculture and free trade as the cornerstones of food security, environmental preservation and mutual cultural respect in a beneficially interdependent manner. Thus, from an ethical perspective, the advantages of intensification and free trade appear consequential.

SUMMARY

Livestock production is undergoing, and is likely to continue to experience, restructuring. Generally the industry is moving toward intensification of production and of international trade affecting distribution. This chapter examines the forces driving this restructuring and explores its impact in three areas of ethics: food security, global environmental protection and cultural diversity.

REFERENCES

Avery, D. (1995) *Saving the Planet with Pesticides and Plastic.* Hudson Institute, Indianapolis, Indiana.

Brookins, C. (1997) *Agricultural Trade Policy.* World Perspectives, Incorporated, Washington DC, March 1997.

Crosson, P. and Anderson, J. (1992) Resources and global food prospects. *World Bank Technical Paper 184.* World Bank, Washington, DC.

FAO (1993) *Production Yearbook*, Vol. 47. Food and Agricultural Organization of the United Nations, Rome, Italy.

Goodland, R.D. *et al.* (1990) The effects of porcine somatotropin and dietary lysine on growth performance and carcass characteristics of finishing. *Journal of Animal Science, Swine* 68, 3261–3276.

Huston, M. (1994) *Biological Diversity.* Cambridge University Press, Cambridge, UK.

International Rice Research Institute (1993) *Rice Research in a Time of Change: a Medium Term Plan for 1994–9.* IRRI, Manila, The Philippines.

McFee, W. (1997) *Testimony of Dr. William McFee, Purdue University Agronomy Department.* Before the US Senate Committee on Agriculture, Nutrition and Forestry, March 1997, Washington, DC.

Ricardo, D. (1965) *The Principles of Political Economy and Taxation, 1817.* Reprint edn 1965. J.M. Dent and Son, London.

Urban, F. and Nightingale, R. (1993) *World Population by Country and Region, 1950–1990 and Projections for 2050.* US Department of Agriculture Economic Research Division, and Food and Agriculture Organization Yearbook, Vol. 47, 1993, Food and Agriculture Organization of the United Nations.

US Foreign Agricultural Service (1998) US exports of rice – calendar year: 1992–96. *Bulk Intermediate and Consumer Oriented (BICO) Commodity Exports.* Washington, DC.

Livestock, Ethics and Quality of Life in Asia: the Food–Feed Dimension of Grain Demand

9

Gurdev S. Khush, Mercedita A. Sombilla and Mahabub Hossain

International Rice Research Institute, Manila, The Philippines

INTRODUCTION

As economic growth proceeds and per capita incomes rise, consumption habits gradually change from a diet primarily based on staple grains to one consisting more of meat, vegetables, fruit, milk and other dairy products. The record high economic growth in many Asian countries has led to significantly higher consumption of beef, pork, poultry, eggs and milk. What are the implications of such dietary shifts on the future demand and supply of grains?

This chapter analyses the food–feed relationship of demand for grains and its implications on future grain supply and on the natural resource base that supports their production. Firstly, we discuss the factors that influence the emerging demand trends for various commodities, particularly the expected shifts in demand composition. This leads to the projection of demand for livestock in the future.

The following section raises the question of whether, in view of the slowing down in production growth in recent years, future grains supply will support the expected growth in demand, not only for food but also for animal feeds. An analysis is given of the possible option of dependence on the international market to fill the widening gap between domestic supply and demand. Some ethical concerns are raised on the relative use of food and feed grains in the light of the need to alleviate poverty and to improve nutritional welfare. The chapter concludes that pressure on the natural resource base will not be eased but will continue in the future as we strive increasingly to grow grains for the expected high demand for both food and feed.

The only bright spot in this otherwise gloomy situation is the emergence of the new science of biotechnology. Biotechnology techniques hold the promise of substantially raising the yield potential of grain crops and of reducing losses caused by diseases and insects. Better quality cereal grains by, for example, changing the amino acid profile and micronutrients could improve nutritional welfare. Therefore, it is important that developing countries invest heavily in biotechnology so that they may produce more and better food and thus reduce their dependence upon imports. In this chapter, metric tonnes are shown by t, million metric tonnes by Mt, million hectares by Mha and tonnes per hectare by t ha^{-1}.

POPULATION AND INCOME GROWTH EFFECT ON DEMAND FOR FOOD

Population and income growth are two major factors that affect demand for grains. Table 9.1 shows that the population of Asia is projected to grow at the rate of 1.1% per annum from 3.1 billion in 1995 to about 4.1 billion in 2020. While this growth rate is less than in the previous three decades, it will still mean an addition of about a billion people to the region, primarily due to the current huge population base. China and India are key examples. Despite the significant reduction in their projected growth rates to about 0.7 and 1.3%, respectively, the combined increment in the population of these two countries between 1995 and 2020 will account for 62% of the Asian population. Demand for food in the region is therefore expected to increase significantly based on mere population growth alone.

Rapid income growth, on the other hand, has greatly influenced shifts in dietary patterns from one which consists mainly of the staple food to one which is more diverse, with a growing contribution of high-value items particularly meat and other dairy products. Economic theory says that as per capita income increases, demand for the staple foods such as cereals tends to decrease. This is evident in Table 9.2 which shows past trends of the per capita cereal and meat consumption in Asian countries. Bangladesh, India, Pakistan, Myanmar, Vietnam, Indonesia and The Philippines, which are categorized as low-income, poverty-stricken countries, still exhibiting either a constant or increasing level of per capita demand for cereals, while the rest of the Asian economies that belong to the upper middle and high income categories show declining per capita demand for cereals. In contrast, per capita demand for meat in most countries is on the rise and has been most rapid in South Korea, China, Japan and Malaysia.

The income elasticity parameters (or the percentage change in per capita food demand for a percentage change in income) in Asian countries for various commodities are shown in Table 9.3. As shown in the table, income elasticities of cereals are small and for some countries are even negative (e.g. Japan and South Korea for rice and several countries for maize).

Table 9.1. Population and income characteristics of the Asian regio

Country region	Population (× 10³)			Population growth rate (%) 1995–2020	Inc p cap (US, 1995	...ιc (%) 1990–1995
	1995	2010	2020			
South Asia	1244.6	1592.7	1796.15	1.48	350	4.6
India	929.0	1152.3	1271.6	1.26	340	4.6
Pakistan	136.3	200.6	247.8	2.42	460	4.6
Bangladesh	118.2	151.9	171.4	1.50	240	4.1
Nepal	21.5	30.7	37.5	2.25	200	5.1
Sri Lanka	17.9	21.0	23.1	1.03	700	4.8
Southeast Asia	481.9	594.9	660.7	1.27	800[a]	10.3[a]
Indonesia	197.5	239.4	263.8	1.17	980	7.6
Thailand	58.2	64.6	67.8	0.61	2740	8.4
Malaysia	20.1	26.2	29.8	1.58	3890	8.7
The Philippines	67.8	88.8	99.9	1.56	1050	2.3
Vietnam	73.8	92.3	104.2	1.39	240	8.3
Myanmar	45.1	57.5	64.3	1.43	—	6.1[b]
Laos	5.0	7.8	10.5	3.00	350	6.5
Cambodia	10.0	15.1	19.9	2.80	270	6.4
East Asia	1421.3	1578.9	1664.2	0.63		
Japan	125.1	127.0	123.8	−0.04	39,640	1.0
China	1220.2	1365.0	1448.8	0.69	620	12.8
South Korea	44.9	50.0	51.9	0.58	9700	7.2
Asia	3147.8	3766.5	4121.1	1.08	—	—
World, total	5687.1	6890.8	7671.9	1.20	4880	2.0

[a]East and Southeast Asia combined.
[b]1996 figure.
Source: UN (1997), ADB (1997), World Bank (1997).

This indicates that, as per capita income grows further, per capita demand for the commodity declines. Income elasticities for livestock products, on the other hand, are positive and relatively large, ranging from 0.5 to 0.7. This fact explains the significant expansion of livestock consumption in the last three decades, particularly in East and Southeast Asia where economic growth has been fastest. Between 1965 and 1995, demand for meat products increased more than fourfold from 2.6 Mt in 1965 to 10.7 Mt in 1995 (FAO, 1997). Even countries in South Asia which are primarily vegetable based have shown increases in demand for meat to the tune of about 4.1 Mt during the same time period. Among meats, pork dominates in most of Asia, especially in areas of Chinese influence, but beef is highly valued in Japan and South Korea. Demand for poultry meat is increasing throughout the region. Egg and dairy product consumption has risen in almost all countries.

. able 9.2. Trend in per capita consumption (kg per capita) of cereals and meat products.

	Cereals			Meat		
	1965	1995	Growth rates (%)	1965	1995	Growth rates (%)
Bangladesh	173	170	−0.04	3	3	0.00
India	146	163	0.39	4	4	0.00
Pakistan	138	149	0.26	7	14	2.34
Indonesia	103	184	1.94	4	10	3.10
Thailand	150	121	−0.73	13	17	0.90
Malaysia	155	126	−0.68	14	50	4.33
The Philippines	117	137	0.53	15	26	1.85
Vietnam	168	181	0.25	11	19	1.84
Myanmar	146	221	1.38	6	8	0.96
China	150	129	−0.51	9	44	5.43
Japan	151	129	−0.52	11	44	4.73
South Korea	184	168	−0.30	5	39	7.09

Source: FAO (1997).

Table 9.3. Food demand response to changes in per capita income for various commodities (income elasticity parameters).

	Beef	Pork	Lamb	Poultry	Eggs	Milk	Wheat	Maize	Rice
India	0.63	0.58	0.38	0.76	0.35	0.75	0.22	−0.01	0.14
Pakistan	0.6	0.22	0.28	0.76	0.4	0.7	0.2	−0.1	0.18
Bangladesh	0.55	0.1	0.38	0.96	0.64	0.75	0.28	0.1	0.21
Indonesia	0.75	0.72	0.28	0.7	0.55	0.65	0.36	0.05	0.15
Thailand	0.77	0.65	0.26	0.68	0.6	0.54	0.29	−0.07	0.04
Malaysia	0.85	0.6	0.28	0.5	0.51	0.38	0.3	−0.15	0.06
The Philippines	0.82	0.72	0.31	0.9	0.5	0.7	0.33	−0.05	0.21
Vietnam	0.82	0.65	0.36	0.89	0.53	0.7	0.35	0.05	0.15
Myanmar	0.82	0.62	0.36	0.94	0.7	0.5	0.34	0.05	0.15
China	0.8	0.64	0.36	0.74	0.51	0.6	0.23	−0.07	0.1
Japan	0.33	0.33	0.13	0.53	0.56	0.1	0.25	−0.18	−0.05
South Korea	0.79	0.85	0.53	0.98	0.83	0.6	0.32	−0.2	−0.1

Source: Rosegrant *et al.* (1995).

Asia, as a whole, has thus emerged as a major importer of meats, reaching about 28% of global meat imports in 1995 (Table 9.4). A huge part of the increment in world meat imports is explained by the East and Southeast Asian countries. No other regions surpass such a tremendous growth in imports.

Table 9.4. Increase in meat imports, selected regions, 1965 and 1995 (Mt).

	1965	1995	% Change
World	4.7	18.6	396
Developed, all	4.3	14.3	333
Developing, all	0.3	4.3	1433
Latin America (incl. Caribbean)	0.129	0.968	750
Sub-Saharan Africa	0.051	0.257	504
Asia, all	0.231	5.164	225
East and Southeast Asia	0.072	1.503	2088
South Asia	0.001	0.005	500

Source: FAO (1997).

Future demand for livestock products

Recent projections show that world demand for meat will grow at the rate of 1.8% per annum between 1993 and 2020 (Table 9.5). Most of this growth will be accounted for by the developing countries. Growth in demand in the developed world will continue to slow down because of stabilized populations, a reduction in the income elasticity of demand for meat as people become more health conscious, and the slowed rate of growth in per capita income. Asian countries' demand for meat products will grow at 3.2% and will thus account for most of the global increases in demand. This higher demand will come from continued rapid growth in the population in most areas as well as income growth, both of which are assumed to be sustained at their current high levels (Table 9.2). The same picture is projected for the world demand for eggs and milk; growth in demand for these products will be about 1.6 and 1.1%, respectively, most of which again will be accounted for by the Asian region.

In volume terms, demand for meat in Asia will increase from 60.8 Mt in 1993 to 135.7 Mt in 2020, or an increase of about 75 Mt (Table 9.6). More

Table 9.5. Projected growth in demand for livestock products (%).

	1993–2020		
	Meat	Eggs	Milk
World	1.82	1.6	1.06
Developed	0.46	0.4	0.18
Developing	2.93	2.31	2.38
Latin America	2.15	1.86	1.21
Sub-Saharan Africa	3.39	2.96	1.51
WANA[a]	2.67	2.24	1.14
Asia	3.17	2.39	3.2

[a]West Africa and North Africa.
Source: Rosegrant *et al.* (1995).

Table 9.6. Projected supply, demand and net trade of meat, milk and eggs (× 1000 t), 1993–2020.

Country regions	All meat 1993			All meat 2020			Milk 1993	Milk 2020	Eggs 1993			Eggs 2020		
	Production	Demand	Net trade	Production	Demand	Net trade	Demand	Demand	Production	Demand	Net trade	Production	Demand	Net trade
Asia	59,374	60,777	−1,403	128,476	135,676	−7,200	101,748	229,834	19,576	19,552	24	2,962	2,919	43
India	3,948	3,838	110	8,041	8,481	−439	60,883	151,592	1,354	1,349	5	2,583	2,534	49
Pakistan	1,572	1,576	−4	3,361	3,829	−468	17,135	40,803	253	252	1	533	593	−61
Bangladesh	334	334	0	547	766	−219	1,804	2,575	98	98	0	213	225	−12
Other S Asia	223	220	3	311	514	−203	1,237	2,506	67	67	0	146	182	−35
Indonesia	1,619	1,722	−103	4,556	4,790	−232	662	977	568	569	−1	1,281	1,280	1
Thailand	1400	1,222	178	3,130	3,197	−66	202	263	535	530	5	1,111	1,033	78
Malaysia	992	898	94	1,954	2,140	−186	43	52	364	359	5	717	723	−6
The Philippines	1,410	1,457	−47	3,256	3,874	−617	32	47	334	334	0	676	703	−27
Vietnam	1,223	1,213	10	2,562	3,002	−441	69	183	197	191	6	363	423	−60
Myanmar	302	302	0	586	728	−142	539	1,118	46	46	0	85	97	−13
Other SE Asia	178	170	8	313	399	−87	29	55	49	49	0	90	108	−17
China	39,403	38,553	850	89,143	89,425	−282	8,015	15,079	12,262	12,261	1	22,431	22,392	39
Japan	3,247	5,025	−1,778	3,849	6,063	−2,215	8,530	11,913	2,584	2,584	0			
Other E Asia	3,523	4,247	−724	6,867	8,469	−1,602	2,568	2,671	865	863	2	1,699	1,776	−77

Source: Rosegrant et al. (1995).

than 50% of this increase will be accounted for by pork, mostly from China. However, the rate of growth of demand for poultry will be fastest, followed closely by beef. Demand for milk will also increase at a relatively fast rate, especially in South Asia.

As shown in Table 9.2, the present levels of per capita consumption of meat are very low. Demand is less than 15 kg per capita per year in South Asia, Indonesia and Myanmar, 17–26 kg in Vietnam, The Philippines and Thailand and between 39 and 44 kg in Japan, China and South Korea. Only in Malaysia is the per capita consumption level relatively high, at about 50 kg per capita, and is expected to continue to grow to about 72 kg per person in 2020 (Table 9.7). China's per capita demand for meat will grow to about 63 kg per person, Thailand to about 47 kg, The Philippines to 39 kg and Vietnam to 29 kg. In Japan, as in many developed countries, per capita demand for meat will slow down, reaching 48.9 kg per capita in 2020.

To meet the projected domestic demand for meat in the region, imports will have to rise to about 7.1 Mt in 2020 as the projected growth in domestic production will only be 3% per annum (Table 9.6). The bulk of the imports will be channelled to the richer East Asian countries which will be able to afford them. However, in most of the other Asian countries, further increases in domestic livestock production would be preferable to save foreign exchange for other purposes. This approach is contrary to the theory of comparative advantage. However, comparative advantage in production cost is not the only issue in Asia. Increasing domestic production would lead to diverting marginal land from the production of rice and wheat to

Table 9.7. Projected per capita demand of meat products in 2020 (kg per capita per year).

	Beef	Pork	Lamb	Poultry	All meat
Asia	5.1	19.4	1.3	7.0	32.6
India	4.1	0.7	0.9	1.0	6.6
Bangladesh	1.8	–	1.0	1.7	4.5
Pakistan	7.7	–	4.9	2.5	15.0
Indonesia	4.1	6.3	0.6	7.1	18.1
Thailand	13.1	11.5	0.0	22.1	46.8
Malaysia	10.3	19.2	0.5	41.8	71.8
The Philippines	5.1	22.0	0.9	10.9	38.8
Vietnam	4.8	18.8	0.1	5.0	28.6
Myanmar	4.1	3.0	0.2	3.8	11.1
China	4.6	45.2	1.4	11.4	62.7
Japan	11.7	16.8	0.5	19.8	48.9
Other East Asia	11.1	35.3	1.7	21.9	69.9
Developed, all	24.9	28.8	3.2	24.0	81.0

Source: Rosegrant *et al.* (1995).

growing of maize, to be used primarily as livestock feed. Taking this course of action, however, raises an important question: will sufficient domestic resources be available to provide feed for a more rapid expansion of the livestock industry? This is not to mention the expected rise in grain demand for food, particularly for rice and wheat, which will still be quite strong in the less developed countries. This is the next issue to be considered.

Projected demand for grains for food and feed

The projected demand for cereal in Asia will increase at the rate of 1.5% per annum between 1993 and 2020. There will be slower growth in demand for food (1.14%) but this will be compensated for by rapid increases in demand for feed, primarily of maize (3.1%). Tables 9.8 and 9.9 show the proportion of demand for food and feed, respectively, to the total demand for cereals in 1993 and in 2020. As can be seen from the table, there will be significant reduction in the food to total demand ratio for maize and other coarse grains but increases in the feed to total demand ratio. For the finer grains, wheat and rice, the food to total demand ratio will remain almost constant. The projected trend is accounted for primarily by continued strong demand for these staples for food, as the poorer segment of the population, which is still huge in many countries, undergoes shifts in their dietary patterns from mostly maize, sorghum, barley and millet, to wheat and rice, as well as to

Table 9.8. Projected proportion of food demand to total demand of cereals.

Country/ regions	Wheat		Maize		Other grains		Rice		Cereals, all	
	1993	2020	1993	2020	1993	2020	1993	2020	1993	2020
India	86.5	85.2	78.4	71.4	87.7	85.8	90.3	90.3	87.9	86.8
Pakistan	91.3	91.3	58.1	53.7	71.3	67.2	89.5	89.3	88.8	88.9
Bangladesh	92.8	92.8	16.7	13.6	93.6	93.5	92.1	92.1	92.1	92.1
Other S Asia	88.9	89.7	82.1	82.2	86.9	86.3	88.5	88.5	87.4	87.7
Indonesia	96.8	96.8	52.6	33.8	100.0	100.0	88.7	88.7	82.9	77.4
Thailand	96.4	96.4	0.8	0.3	27.3	12.1	76.9	77.2	53.4	41.5
Malaysia	71.2	72.7	3.4	1.9	41.6	31.5	95.3	95.3	52.8	49.3
Philippines	99.9	100.0	23.7	13.2	87.0	81.1	88.4	88.4	66.7	60.3
Vietnam	98.8	98.9	64.3	54.1	78.4	68.3	86.0	86.0	85.0	84.5
Myanmar	89.3	89.2	50.0	40.1	82.6	79.7	100.0	100.0	98.6	98.1
Other SE Asia	100.0	100.0	56.8	55.4	40.0	27.3	100.0	100.0	97.6	97.6
China	87.3	82.5	33.2	15.0	66.9	52.3	90.1	90.1	73.0	59.9
Japan	84.3	84.6	17.4	12.6	16.1	11.6	91.7	91.7	46.2	42.1
Other E Asia	58.8	55.9	14.8	8.9	56.4	42.0	93.2	93.3	47.4	37.2
Asia	86.7	84.4	32.9	18.4	69.4	62.9	90.1	90.2	75.7	68.9

Source: Rosegrant *et al.* (1995).

Table 9.9. Projected proportion of animal feed demand to total demand of cereals.

Country regions	Wheat 1993	Wheat 2020	Maize 1993	Maize 2020	Other grains 1993	Other grains 2020	Rice 1993	Rice 2020	Cereals, all 1993	Cereals, all 2020
India	3.2	4.5	10.7	17.6	2.4	4.3	0.4	0.4	2.2	3.3
Pakistan	4.8	4.8	23.1	27.5	20.6	24.7	0.0	0.0	5.7	5.8
Bangladesh	0.0	0.0	50.0	54.5	5.1	5.6	0.0	0.0	0.0	0.1
Other S Asia	8.9	8.2	9.8	9.8	4.1	4.7	1.4	1.4	4.8	4.6
Indonesia	0.0	0.0	41.6	60.5	0.0	0.0	2.0	2.0	8.8	14.6
Thailand	0.0	0.0	98.4	98.8	47.7	62.9	4.0	4.0	33.6	48.4
Malaysia	14.0	12.6	92.2	93.7	31.2	42.3	2.0	2.0	41.1	44.1
The Philippines	0.0	0.0	73.3	83.8	9.7	15.9	7.5	7.5	30.2	36.7
Vietnam	0.0	0.0	28.9	39.1	18.9	26.7	1.2	1.2	2.8	3.5
Myanmar	0.0	0.0	47.9	58.0	8.7	11.4	0.0	0.0	1.1	1.5
Other SE Asia	0.0	0.0	25.9	26.6	40.0	36.4	0.0	0.0	1.4	1.4
China	5.7	10.6	66.6	84.8	19.4	34.0	1.7	1.7	21.1	34.8
Japan	8.3	8.0	72.8	77.5	80.2	84.7	0.0	0.0	45.8	50.0
Other E Asia	23.1	26.1	61.1	68.2	27.2	39.6	0.0	0.0	34.8	44.4
Asia	5.5	7.9	62.2	77.3	20.0	26.4	1.3	1.3	16.6	23.8

Source: Rosegrant *et al.* (1995).

meat and other high-value products. Significant increases in per capita demand for cereals in some of these countries are not expected since they are already relatively high. These will either be maintained at their current levels or will rise very slightly.

Unlike cereals, average per capita demand for livestock products in Asia is still far below the level of the more developed countries (Table 9.2). Projection estimates also show Asia trailing behind the more developed regions in terms of per capita levels of demand for meat (Table 9.7). However, some countries have a strong potential for increasing their demand for meat at rates which are higher even than the projected rates. China, for example, which has consistently shown double-digit income growth rates in the past decade, could very well stage a more dramatic shift in demand such that per capita meat consumption will grow much faster, to reach the level projected for Malaysia. If such a phenomenon takes place, China's demand for meat in 2020 would reach 102.7 Mt, 13 Mt more than the level indicated in the current projection estimates. A similar scenario is not unlikely for other Asian countries.

How will these possibilities translate into demand for grains? At the current feed to meat conversion ratio of about 3.95 kg in China, an additional cereal requirement of about 51 Mt will be needed in 2020 to increase livestock production to the level that would meet the increments in domestic demand for meat. Such additional grain requirements will subsume about 20% of the projected world supply of cereals and will put great pressure on the world cereal market, especially for maize which is the principal grain for animal feeds. In addition, further expansion and intensification of grain-fed livestock especially pigs and poultry, which is advancing rapidly in much of Asia, is accompanied by increases in feed to meat conversion ratios. Thus, the 3.95 kg of feed needed to produce 1 kg of meat in China could rise. This will mean additional pressure on the grain market unless these countries achieve some degree of technical efficiency in animal feeding. Otherwise Governments will have to curtail consumption of meat in order to maintain the availability of food grains at reasonable prices particularly for the segment of society on a lower level of income.

EMERGING TRENDS IN THE GLOBAL CEREAL MARKET

This section focuses on rice and corn; on rice primarily because it is the major staple food in Asia, and on corn because it is a major ingredient in animal feed.

About 90% of the world rice supply comes from Asia. Between 1965 and 1975, rice production grew at 2.9% per year with significant contributions from growth in the area used for production (Fig. 9.1). Rice production strengthened further during the following decade at 3.2% per annum, most of which then came from yield contributions. During 1985–1994, growth in rice production was only 1.6% per year.

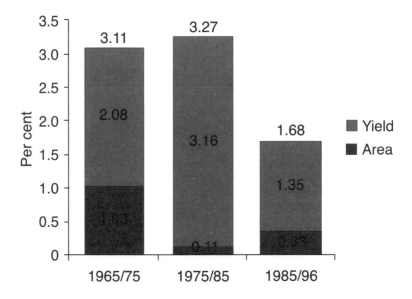

Fig. 9.1. Area and yield contribution to rice production in Asia, 1965–1996. (FAO, 1997).

Table 9.10. Trends in domestic production of corn in selected countries/regions, 1962–1994.

Country	Crop area (× 1000 ha)			Yield (t ha^{-1})			Production (Mt)		
	1962– 1964	1982– 1984	1992– 1994	1962– 1964	1982– 1984	1992– 1994	1962– 1964	1982– 1984	1992– 1994
USA	22,975	26,452	28,047	4.09	6.98	7.76	94.1	208.0	219.4
Latin America	21,717	25,798	27,636	1.22	1.89	2.47	26.4	48.8	68.3
Africa (developing)	11,797	14,493	17,516	0.99	1.26	3.99	11.7	18.3	24.5
China	14,522	18,674	20,821	1.37	3.62	4.84	19.9	67.5	100.8
Indonesia	3,130	2,716	3,203	1.00	1.66	2.19	3.1	4.5	7.0
Thailand	416	1,923	3,218	1.97	2.33	2.96	0.8	3.8	3.6
The Philippines	1,923	3,218	3,240	0.67	1.03	1.52	1.3	3.3	4.9
World	106,659	122,990	130,309	2.00	3.37	4.01	213.5	415.0	522.6

Source: FAO (1997).

A similar scenario exists for corn, except that the major player is the USA from where more than 50% of the corn supply comes (Table 9.10). Expansion of the cultivated area was a major factor behind the growth in US corn production during the 1962–1984 period. In recent years, the harvested area has stabilized at 28 Mha. There has been a spectacular expansion of yield with rapid technological progress from 4.1 t ha^{-1} in 1962–1964 to 7.9 t ha^{-1} in 1994–1996. Further increases in yield would be difficult. The growth in production has slowed from 3% per year during 1962–1984 to only 0.9% during the last decade. This trend can also reduce the growth in world production in the long term. Another major source of corn supply is Latin America. About two-thirds of all maize produced in the developing world (except China) comes from these regions. Asia's share of world corn production is now about 25%, a modest increase from the 17% in the early 1960s. China, Thailand, Indonesia and The Philippines are the major contributors to the region's corn supply.

However, production growth in these countries is not expected to increase significantly. As with rice, most of these countries show a reduction in growth rates during the last decade compared with the earlier period as the potential for further increasing yield levels is also reaching its limit. There are several reasons for the observed deceleration of growth, namely an increasing scarcity of key inputs for production such as land and water, the difficulty in increasing farm profitability and the rapid approach of maximum potential yield levels that current technologies offer. These are discussed in more detail in the next section.

FACTORS INFLUENCING GRAIN PRODUCTION IN ASIA

Increasing scarcity of land and water

Natural resource constraints to increasing rice production are becoming severe for most of the low-income countries in Asia. As the frontier of cultivable land was closed long ago, the per capita availability of arable land has been declining rapidly with the growing population. The rice-growing regions of China now support 17 persons per hectare of arable land; the figure is 13 for Bangladesh, 11 for Vietnam and 8–10 for India, Indonesia and The Philippines. Only Thailand, Myanmar and Cambodia have favourable endowments of land, with two to four persons per hectare. The population pressure is reflected by the high cropping intensity for food grain production, and heavy use of fertilizers and pesticides per unit of land.

Most observers agree that the area under rice cultivation is expected to decline with economic prosperity and urbanization, as the demand for land for non-agricultural uses grows. There will also be economic pressure to release rice land in favour of producing vegetables, fruit and fodder, whose market becomes stronger with economic progress. In China, the rice harvested area declined from 37 Mha in 1976 to 31 Mha in 1996; in The Philippines, it declined from 3.7 to 3.2 Mha within the same period. In Java, Indonesia, nearly 50,000 ha of land are taken out of rice cultivation every year. To compensate for this loss, the government plans to open up environmentally fragile coastal wetlands.

Water, which is usually regarded as an abundant resource in humid Asia, is also becoming a scarce commodity. In absolute terms, annual water withdrawals are by far the greatest in Asia, where agriculture accounts for 86% of total annual withdrawal compared with 38% in Europe and 49% in North and Central America. The per capita availability of water resources declined by 40–60% in most Asian countries over the 1955–1990 period and are expected to decline further due to continued population growth (Gleick, 1993). As water shortage looms on the horizon, many governments may adopt policies to reduce water consumption in rice cultivation to meet the growing demand for non-agricultural uses.

The scope for further conversion of rain-fed land to irrigated land, which was the major source of past production growth, is also becoming limited. The cost of irrigation has increased substantially, as easy options for irrigation development have already been exploited (Rosegrant and Svendsen, 1993). Also, environmental concerns have been growing regarding the adverse effects of irrigation and flood control projects on waterlogging, salinity, fish production and the quality of ground water. Already, there has been a drastic decline in investment for the development and maintenance of large-scale irrigation projects in many Asian countries.

Sustaining incentives in rice farming

Even if all the physical and environmental constraints to production growth are overcome, many countries of South and Southeast Asia will still face the problem of sustaining farmers' interest in rice cultivation as they continue to prosper. The expansion of the non-farm sector and the rapidly rising labour productivity have contributed to a long-term upward trend in rates of pay in urban centres. Prospects of higher lifetime earnings and improved living conditions promoted migration of labour from rural areas to cities, which in turn pushed up agricultural wages. Agricultural rates of pay increased faster in countries with faster economic progress and higher levels of income (Table 9.11). Since traditional rice farming is a highly labour-intensive activity, the increase in wages inflates the cost of rice production and reduces profits and farmers' incomes. In Japan and South Korea, the constant outflow of the agricultural labour force has caused a continuous decline in the farming population. As the age of workers and depopulation in remote areas have continued to increase it is difficult to sustain the existing rural communities in some areas.

With the growing scarcity of land, labour and water, and the upward trend in their prices, the cost of rice production has continued to increase in spite of more efficient use of inputs (through improved crop management practices) and labour saving (through mechanization in rice cultivation). In 1987, the cost of rice production was 20 times higher in Japan and seven times higher in South Korea than in Thailand (Table 9.12). As competition for scarce inputs grow with the development of the non-farm sector, rice farming becomes uneconomic. More and more farm household members take up non-farm employment and become part-time rice farmers. Under political pressure from farm lobbies, the government had to protect the domestic rice market, in order to increase prices, and provide farm subsidies, in order to keep the balance in incomes of rice farmers and urban labour households. Rice production in Japan and South Korea had been

Table 9.11. The relationship between economic prosperity and agricultural wage.

Country	Per capita income (US$, 1994)	Agricultural wage (US$ day^{-1})	
		1966	1991
Bangladesh	220	0.63	1.39
The Philippines	950	0.74	2.56
Thailand	2,410	0.48	2.51
South Korea	8,260	0.95	33.30
Japan	34,630	2.50	52.00

Source: IRRI (1995).

Table 9.12. Costs of production and farm gate prices of
paddy rice in selected countries, 1987–1989.

Country	Cost of production (US$ t^{-1})	Farm gate price (US$ t^{-1})	Paddy yield (t ha^{-1})
Japan	1987	1730	6.5
South Korea	939	957	6.6
USA	195	167	6.3
Vietnam	100	130	4.6
Thailand	120	141	1.8
Bangladesh	138	180	2.7

Sources: IRRI (1995).

adjusted to domestic demand through manipulation of trade, pricing and
subsidy policies.

Approaching technological limits

Recent developments indicate stagnation of yield at high-income levels for
major grains. Rice yield, for example, has remained stagnant at around
6.5 t ha^{-1} in Japan and South Korea after reaching that level in the late
1960s and late 1970s, respectively (Table 9.12). Countries in South and
Southeast Asia are starting to show the same pattern of yield stagnation
although at lower levels of yield. Increased pest pressure and frequent
cloudy days with below optimal sunshine hinder many of these countries in
attaining the higher yield levels reached by the more temperate areas
(Seshu and Cady, 1984; Seshu, 1988). Recent studies likewise have shown
that the best farmers' yields are now approaching the potentials that
scientists are able to obtain in their experimental fields with today's knowledge (Pingali *et al.*, 1997). The emerging trend seems to indicate that the
future prospect is for stagnation and/or decline in rice production.

There are several factors that account for deterioration in rice productivity. It should be noted that the impressive performance of the current
technology in the past three decades was made possible by the combined
effect of irrigated land expansion, the adoption of modern rice varieties in
these areas and the application of fertilizer. The gradual exhaustion of these
frontiers in the last three decades has led to the general slowing down of
productivity growth in recent years. Most rice farms are now irrigated and
planted to modern rice varieties. In most of the irrigated areas, fertilizer use
is already optimal (David and Otsuka, 1994). All these allowed the intensive
monoculture of rice in irrigated lands, the long-term effect of which led to
deterioration in rapid soil and water quality making it difficult for farmers to
sustain high yields (Flinn and De Datta, 1984; Cassman and Pingali, 1995a).

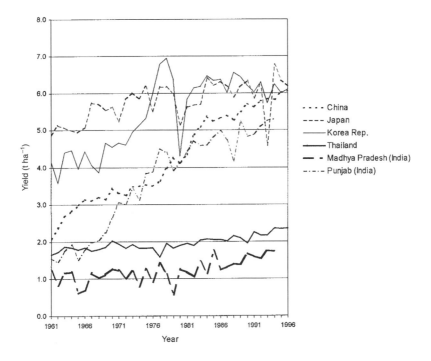

Fig. 9.2. Trends in rice yield in selected Asian countries, 1961–1996.
(IRRI, 1995).

Similar yield growth contraction has also been noted in maize. China, which had a buoyant yield growth at 4.8% per year, has been experiencing a stagnant yield in recent years. Exploiting maize technology potential could still continue to increase yields in the region but probably not sufficiently to meet the expected increase in demand for the commodity. In Thailand, strong yield growth in maize was achieved in recent years with more widespread use of high-yielding seed varieties. This somehow overcame the downward effect of significant decline in maize areas because of commercialization and rapid degradation of lands. Yield increases have been of major importance in increasing output in Indonesia, especially in 1962–1984, and recently in The Philippines. In both countries, the crop is produced in upland areas and on hill slopes under rain-fed conditions. The majority of the maize areas in these countries are still planted to early developed and local varieties. The adoption of high-yielding hybrid seed varieties is still low. There is potential for expansion of yield particularly if irrigation facilities become available, but this is unlikely to replicate the spectacular achievements of the past for rice and wheat.

Impact of liberalization in rice trade

The implementation of the recently concluded GATT–Uruguay Round negotiations may dampen further incentives for rice production, particularly in middle and high income countries (Pingali, 1995). These countries will not be able to compete with the low-income economies whose cost of labour is low, or with large land-surplus countries in the developed world (e.g. Australia and the USA) who reap economies of scale because of the large size of rice farms. If the domestic market is opened for competition, the price of rice will decline substantially, providing incentives for consumers to choose imported food staples, and forcing farmers to abandon rice cultivation in favour of more lucrative economic activities.

An important way of gaining competitive strength in the face of liberalization of rice trade is consolidation of smallholdings into large-scale farms, as rural households migrate to urban areas leaving their land behind. The 'smart farming' in large-scale holdings, as currently practiced in the developed world, may contribute to the vertical integration of the rice industry, more efficient utilization of large-scale machinery and reduction in the large number of part-time farmers who are now involved in the supervision of numerous tiny farms, whose income must be maintained at least at the level of that of urban labour households. The main constraint to the consolidation of holdings into efficient and competitive large-scale farming in Asia is, however, the high price of land that prohibits development of an active land market.

Can we rely on the world market for the region's rice and corn needs?

The international trade in rice has expanded from close to 7 Mt in 1961 to more than 25 Mt in 1997. The rice trade, however, remains very limited, with only about 5% of the world's rice production in the international market. This is in contrast with nearly 20% for wheat and 11% for coarse grains. Thailand, the USA and Vietnam are the more important rice exporters, accounting for about 60% of the export market in the early 1990s (Table 9.13). India emerged with a significant volume of exports in 1994–1995 but failed to sustain this, and Myanmar is starting to regain its position in the export market that it lost in the late 1960s. Among all these exporters, only the USA can be regarded as having the capability for more stable production.

In most Asian countries, rice production is greatly dependent on variable weather conditions that cause periodic shortages and surpluses to occur, producing wide fluctuations in marketable surplus and instability in domestic and international prices. This is not to mention the impact of a rapidly changing economic environment and changes in policy that affect factors of rice production. Current market trends in Thailand indicate that

Table 9.13. World rice market: changes in the pattern of trade, 1961–1993.

Regions	Share of export (%)		Share of import (%)	
	1961	1993	1961	1993
Asia	71.4	65.69	72.9	38.44
East Asia	8.7	9.43	10.3	5.11
Southeast Asia	55.5	44.78	32.1	6.34
South Asia	6.8	10.17	25.8	2.66
Middle East	0.4	1.31	4.7	24.32
Africa	4.6	1	8	24.90
Latin America	4.8	6.58	5.6	14.03
High-income countries	19.2	26.81	13.5	25.33
USA	12.36	16.41	0.09	1.32
Europe	5.87	7.34	11.73	14.08
Volume of trade (Mt, milled rice)	6.8	16.33	6.7	15.36

Source: computed from FAO Statistics (1997).

the country is gradually losing its comparative advantage in rice production and exports, due to the rapid increase in farm wages. Only a decade ago, Thailand taxed rice exports. It has already started providing export sub-sidies to raise prices in the domestic markets in order to sustain farmers' incentives in rice cultivation. Thailand's place as a major rice exporter can easily be taken over by Myanmar, where there is surplus production capa-city that could be mobilized with appropriate and effective policy measures (Hossain and Marlar, 1995). Vietnamese farmers responded favourably to the economic liberalization introduced in recent years (Pingali and Xuan, 1992), and Vietnam became the third most important exporter of rice in the world market due to rapid growth in rice production since the mid 1980s. However, Vietnam has almost exploited its potential for an increase in rice yields and may reduce exports in the future to accommodate the growing internal demand for the commodity.

These all indicate that the future rice market will continue to be small and will remain a hindrance towards the principles of free trade. Unless grain importers have access to commercially available world food supplies, the claims of these countries for the need for self-sufficiency will be difficult to resist. These claims come even from countries highly dependent on ex-ternal trade.

World trade in corn is much more extensive than that of rice, with average volume of trade at 14% of total world production in the early 1990s. The proportion has increased substantially from the mere 9% share in the 1960s. More than 90% of world corn exports come from the USA, China, France and Argentina (Table 9.14). This is dominated by the USA, with an average export volume of 42.7 Mt in 1991–1993, about two-thirds of the world total. China came second with its 9.7 Mt of exports. At the same time,

Table 9.14. Major corn importing and exporting countries, 1991–1993.

Country	Quantity × 10³ tonnes	Percentage of world total
World production	495,480	–
World imports	69,136.7	–
Major importers		
Japan	16,630.4	24.1
Russian Fed.	7,227.0	10.5
Korea	6,098.6	8.8
China	5,430.9	7.9
Spain	1,956.5	2.8
Malaysia	1,692.7	2.4
The Netherlands	1,643.2	2.4
Egypt	1,630.6	2.4
UK	1,552.9	2.2
South Africa	1,525.1	2.2
World exports	69,145.6	–
Major exporters		
USA	42,719.6	61.8
China	9,740.1	14.1
France	6,483.7	9.4
Argentina	4,953.8	7.2

Source: FAO (1997).

however, China is a major importer of the commodity, with average imports of about 5.4 Mt during the period. However, recent trade figures show China becoming a net importer, with net purchases of about 2.5 Mt in 1994–1995 (USDA, 1996). Given the country's huge and growing population as well as expectations of continued rapid economic growth and rising per capita demand, corn imports are likely to increase further as domestic production is incapable of meeting the feed demand of the rapidly expanding livestock sector.

On the other hand, the era of huge purchases in Russia and the newly independent states (NIS) seems to be over. The average import level of about 12 Mt during the 1980s is down to about 7.2 Mt because of smaller domestic demand resulting from economic adjustments and market reforms. Other major Asian importers include Japan, South Korea and Malaysia; their aggregate imports account for more than one-third of the world total.

The world corn market, however, is a more stable market compared with that of rice because of the existence of reliable suppliers such as the USA. This stability will be strengthened as the implementation of the Uruguay Round agreement proceeds with more prompt and accurate international price signals reflecting supply and demand balances in different countries.

The other condition for dependence on the world market is the availability of sufficient foreign exchange for import purchases. In this regard, countries should achieve favourable economic growth in order to obtain sufficient foreign exchange. At the same time, households should have productive employment and adequate incomes to allow them to acquire the required food from the market. Countries in East and Southeast Asia are fortunate in this respect, especially those that have not been very much affected by the recent financial crisis. Depending on how fast they recover, The Philippines and Indonesia may have to strive to increase rice production domestically to save the use of foreign exchange for some other import needs. This holds true for the other low-income economies.

ETHICAL CONCERNS

Several developments in the production and distribution of food and feed grains raise ethical concerns such as:

- increasing diversion of land from food to feed grains;
- pressure on increasing yield through greater use of agrochemicals; and
- increasing food grain prices due to liberalization of trade in staple grains.

An important ethical issue is the existence of massive poverty and food insecurity in spite of the impressive growth in production achieved since the green revolution was initiated in the mid 1960s. Poverty, a dominant feature of life in a large number of developing countries, afflicts a total of about one billion rural people in the world (World Bank, 1990; Jazairy and Alamgir, 1992). While urban poverty is growing, a more serious concern is the persistent and widespread rural poverty. The rural poor account for more than four-fifths of the number of urban poor. In Asia, with about 60% of the world's population and 78% of the world's rural population, about one-third of the rural population lives in poverty. In a situation where many countries and population groups are still far from having met their needs for direct food consumption of cereals, it is valid to ask: is it morally justified to divert land and other resources now tied up in the production of food grains to growing of livestock feed, vegetables and fruit, to meet the growing demand for these products by the middle and high-income groups? The diversion of inputs and the consequent negative impact on the growth of food grain production may ultimately raise food grain prices or halt the decline in real prices (adjusting for inflation) which will affect the purchasing capacity and the entitlement to food for the poor (Sen, 1981).

Another important ethical concern is the sustainability of food grain production due to over exploitation of natural resources which will affect the food availability of future generations. Much of the increasing supply in food grain production, particularly in Asia, has come from the intensification of production on existing land, and an increasing yield from excessive use of chemical fertilizers and spraying of harmful pesticides that farmers employ to reduce the yield losses from pests. In Vietnam, Bangladesh and Java, Indonesia, where population pressure is very high, the cropping intensity in arable land has reached almost 200%. It is reported that these practices have led to a decline in the capacity of soils to supply nitrogen, a deterioration in water quality, a building up of toxic chemicals in food, and exposure to health hazards associated with the spraying of pesticides (Brown, 1990; Rola and Pingali, 1993; Cassman and Pingali, 1995b; Pingali *et al.*, 1997).

The diversion of prime agricultural land to non-agricultural uses to accommodate growing urbanization is putting further pressure on agricultural intensification and increasing use of agrochemicals. In view of these developments, agricultural scientists must show ways to economize on the use of agrochemicals, explore developing durable resistance in improved seeds against insect pests and diseases that reduce farmers' need to use agrochemicals, and shift the yield potentials of food grains to economize on the use of scarce land.

Developed countries of the world with grain surpluses (such as Australia, Canada and the USA) are exerting pressure to liberalize international trade in food grains in order to gain access to potential import markets in the developing countries. This attitude also raises an ethical issue. Many countries in the developing world have followed a strategy of sustaining food security through self-sufficiency in domestic production of staple grains. In pursuance of this strategy, they have created a protective wall in the domestic market in order to maintain farmers' incentive in production when wages and land prices increased due to growing competition for these inputs from the high-productive non-farm sectors of the economy. This has led to a maintenance high-cost domestic food grain production industry. For example, in Japan and Korea, the cost of domestic production of food grains is 10–20 times higher than the cost of production in the USA, Thailand and Vietnam (Yap, 1991). If these countries are exposed to competition following the liberalization of international trade, the food grain production sector may not survive. If improving economic efficiency is the primary consideration, the strategy of acquiring food from countries that produce at a lower cost rather than producing it at a high cost in the domestic market, makes economic sense.

However, from an ethical viewpoint, it is important to take a dynamic view of the issue. What will happen if every natural resource-scarce country abandons the production of staple grains in order to release land and labour to more profitable economic activities, and opts for sustaining food security through international trade? What would happen to total supplies of food grains to the world market if the domestic food grain industry in the importing countries collapses? What would happen to the prices of food grains in the international market if the fall in production in the importing countries cannot be compensated for by an increase in supplies from the exporting countries?

Provided there is free trade, it is *not* difficult for high-income food-deficit countries and the affluent consumers to access their staple food from the market, even when there is a scarcity. The market will distribute the scarce supplies in favour of the affluent who can pay higher prices. It is the poor consumers in the low-income poverty-stricken countries who will suffer when there is a scarcity of staple food. When prices soar, the government may intervene in the market to protect the interest of the nation. Imposing a ban on exports of staple food when there is a scarcity in the domestic market is *not* a rare phenomenon. Also food-exporting countries sometimes use food aid to interfere in the domestic politics of countries who are in dire need of accessing food from the world market.

In view of the above considerations, it is ethical for the world community to allow each country to produce the staple food that they need, rather than making them dependent on others in order to meet the most basic human need.

CONCLUSIONS

The above analysis shows that economic growth and rising incomes are not barometers for complacency in grain production. While on the one hand they tend to shift diets away from the staple foods and reduce the per capita demand for grain, their derived demand in terms of animal feeds would easily make up for these losses. In a region such as Asia where, due to great inequality in income distribution, a large portion of the population are still food insecure, a sufficient grain supply availability should be ensured. Pressure on land and other natural resources will therefore continue in the future as grain production is intensified to meet increased demand for food in the low-income economies and for animal feed in the medium to high-income economies.

A number of social and economic policy measures are needed to facilitate application of science for sustainable food and nutrition security. There is an urgent need for the adoption of population policies in Asia. The spread of democratic systems of governance is necessary in order to introduce land reform measures, security of land tenure and minimum wages for farm workers. Policies for curbing population growth and agrarian reform in developing countries and for sustainable lifestyles in industrialized countries are vital for achieving a hunger-free world.

Fifty years of unprecedented gains in crop and livestock productivity, in particular the fruits of the green revolution, have given the world an essential breathing space to prepare for even greater challenges of the future. Science is now poised to take the next step in close cooperation with policy makers. Innovations should be equally applicable to small and large farmers, including those emerging from the fields of biotechnology, agrometeorology and information technology, as well as holistic management systems of soil health care, judicious water use, integrated pest management and integrated nutrient management. These innovations are only a few of the approaches needed to reach 500 million undernourished Asians. In addition, growing opportunities exist for rural agroprocessing and agribusiness enterprises to take advantage of the changing consumption patterns of an increasingly urban world population.

Tapping this potential will depend upon strengthening the capacity of national agricultural research and development systems to respond to these new challenges with creativity. Therefore, the governments in Asia should reverse the global trend of disinvestment in agricultural research and development. Meeting the challenge of increasing food availability now and in future demands equal focus on production systems and on the issues of access to food such as employment opportunities. This will require a number of innovative approaches.

- Production of more grains from a diminishing resource base, requiring new agricultural technologies and management systems, providing increased productivity per unit of land, water, energy, labour, time and investment.
- A systems approach marshalling the combined and coordinated efforts of physical scientists, agricultural researchers and social scientists, while agricultural production, including livestock, will remain the foundation of food security. The larger scientific framework must integrate distribution systems, rural development, as well as economic and social empowerment of the poor, including women.
- National policies for sustainable food and nutrition security should ensure that every individual has physical, economic, social and environmental access to a balanced diet, safe drinking water, sanitation, environmental hygiene, primary health care and a productive life. Food should be produced with efficient and environmentally benign technologies that conserve and enhance the natural resource base of crops, animal husbandry and forestry.

Developed countries should consider reversing the downward trend of development assistance to poor countries. Failure to make ethically correct decisions will result in continued low economic growth and rapidly increasing food insecurity and malnutrition, lost opportunities for expanded international trade, widespread conflict and civil strife, which will make the world insecure and unstable for all.

SUMMARY

The present Asian population of 3.1 billion is projected to grow at the rate of 1.1% per annum, reaching about 4.1 billion in 2020. Continued rapid economic growth in Asia has greatly influenced dietary patterns, from one which consists mainly of the staple food to one which is more diverse and which has a growing component of high-value items, particularly meat and dairy products. As a result, Asia has emerged as a major importer of meats, and the region's share of global meat imports reached 28% in 1995. Recent projections show that the demand for meat in Asia will grow at 3.2% per year. This increased demand will be a consequence of two factors: continued rapid growth in population in most Asian countries and income growth. In volume terms, demand for meat in Asia will increase from 60.7 Mt in 1993 to 135.7 Mt in 2020. The richer countries of East Asia will be able to afford imports of meat, but most other Asian countries will have to increase domestic livestock production, thus further increasing the demand for feed grains, particularly corn.

Recent trends in production of both corn and rice (a staple in Asia) raise serious concerns about the sustainability of growth in production to meet projected demands. As a consequence, Asian countries must increase investment in agricultural research to develop technologies for raising productivity without adversely affecting the resource base. This will require difficult decisions to reduce spending in some other areas such as armaments. Developed countries should consider reversing the downward trend of development assistance to poor countries.

REFERENCES

ADB (1997) *Key Indicators for Asian and the Pacific Region.* Asian Development Bank, Manila, Philippines.

Brown, L.R. (1990) *State of the World.* World Watch Institute, Washington DC.

Cassman, K.G. and Pingali, P.L. (1995a) Extrapolating trends from long-term experiments to farmers' fields: the case of irrigated rice systems in Asia. In: Barnet, V., Payne, R. and Steiner, R. (eds), *Agricultural Sustainability: Economic, Environmental and Statistical Considerations.* John Wiley & Sons, London; pp. 63–84.

Cassman, K.G. and Pingali, P.L. (1995b) Intensification of irrigated rice systems: learning from the past to meet future challenges. *Geojournal* 35, 299–306.

David, C.C. and Otsuka, K. (1994) *Modern Rice Technology and Income Distribution in Asia.* Lynne Rienner Publishers, Boulder and London.

FAO (1997) *FAO Statistics.* Food and Agriculture Organization, Rome, Italy.

Flinn, J.C. and De Datta, S.K. (1984) Trends in irrigated rice yields under intensive cropping at Philippine research stations. *Field Crops Research* 9; 1–15.

Gleick, P.H. (1993) About the data. In: Gleick, P.H. (ed.), *Water Crisis: a Guide to the World's Fresh Water Resources.* Oxford University Press, New York, pp. 117–488.

Hossain, M. and Marlar, O. (1995) Myanmar rice economy: policies, performance and prospects. *Journal of Agricultural Economics and Development* 23, 110–137.

IRRI (1995) *World Rice Statistics, 1993–1994.* International Rice Research Institute, Manila, The Philippines.

Jazairy, I. and Alamgir, M. (1992) *The State of World Poverty: an Inquiry into Its Causes and Consequences.* International Fund for Agricultural Development, Rome.

Pingali, P. (1995) GATT and rice: do we have our priorities right? In: *Fragile Lives in Fragile Ecosystems.* International Rice Research Institute, Manila, The Philippines, pp. 25–38.

Pingali, P.L. and Xuan, V.T. (1992) Vietnam: decollectivization and rice productivity growth. *Economic Development and Cultural Change* 49, 697–718.

Pingali, P.L., Hossain, M. and Gerpacio, R.V. (1997) *Asian Rice Bowls: the Returning Crisis.* CAB International, Wallingford, UK.

Rola, A.C. and Pingali, P.L. (1993) *Pesticides, Rice Productivity and Farmers' Health: an Economic Assessment.* World Resources Institute, Washington DC, and IRRI, Los Baños, The Philippines.

Rosegrant, M.W. and Svendsen, M. (1993) Asian food production in the 1990s: irrigation investment and management policy. *Food Policy* 20, 203–223.

Rosegrant, M.W., Agcaoili-Sombilla, M.C. and Perez, N. (1995) *Global Food Projections to 2020: Implications for Investment.* International Food Policy Research Institute, Washington DC.

Sen, A.K. (1981) *Poverty and Famines: an Essay on Entitlement and Deprivation.* Clarendon Press: Oxford.

Seshu, D.V. (1988) Impact of major weather factors on rice production. In: *Workshop on Agrometeorological Information for Planning and Operation in Agriculture,* 22–26 August, Calcutta, India.

Seshu, D.V. and Cady, F.B. (1984) Response of rice to solar radiation and temperature estimated from international yield trials. *Crop Science* 24, 649–654.

UN (1997) *Population Projections, 1996.* United Nations, New York.

USDA (1996) *Economic Outlook Series.* United States Department of Agriculture, Washington DC.

World Bank (1990) *Poverty: World Development Reports, 1990.* Oxford University Press, Oxford.

World Bank (1997) *World Development Report.* Oxford University Press, Oxford.

Yap, C.L. (1991) A comparison of cost of producing rice in selected countries. *Economic and Social Development Papers No. 101.* Food and Agricultural Organization, Rome.

Livestock, Ethics, Quality of Life and Development in Latin America   10

Hugo Li-Pun, Carlos U. Leon-Velarde and Victor M. Mares

International Livestock Research Institute, Addis Ababa, Ethiopia

INTRODUCTION

Latin America is well endowed with natural resources. However, it also has extensive poverty, a slow rate of development, a high population growth rate and unequal distribution of wealth, together with degradation of natural resources in marginal areas.

Over the past decades, policies in the region have promoted the development of large urban centres. As a result, most of the human population is now found in cities. It is estimated that by the year 2025 over 80% of the region's human population will be urban (World Bank, 1992). Industrial development has not kept pace with rural–urban migration, leading to urban unemployment and poverty. While poverty is common in the outskirts of cities, the poorest of the poor are found in rural areas.

The majority of farmers in the region are smallholders. For example, in Central America and the Andean region, smallholders constitute over 80% of the farmer population. Smallholders practice highly diversified agriculture to cope with both climatic and economic risks. Within these diversified farming systems, livestock play critical roles, providing valuable products such as meat, milk, wool, fibre and skins, and inputs such as manure and traction that support sustainable cropping systems. Livestock are also the farmers' preferred form of saving for times of need. In certain ecosystems, such as the semi-arid zones and high-altitude areas, animals have a definite comparative advantage over cropping, as climatic risks (low rainfall, frost or high altitude) preclude safe cropping.

Objectives of this chapter

This chapter addresses four principle aims:

- to highlight some of the physical and socio-economic characteristics of the Latin American region;
- to discuss the economic, social and environmental importance of the livestock sector as a means of overcoming poverty, malnutrition and degradation of natural resources;
- to review ongoing economic and policy trends, how they affect the livestock sector and the implications for equity in access to resources, services and benefits; and
- to propose specific research and development approaches to address some of the major economic, social and environmental challenges to livestock agriculture.

In this chapter, metric tonnes are shown by t, million metric tonnes by Mt, million hectares by Mha, tonnes per hectare by t ha^{-1}.

MAIN CHARACTERISTICS OF THE REGION

According to statistics from the Food and Agriculture Organization of the United Nations (FAO, 1993), Latin America has a surface area of approximately 22 million km^2, 22.7% of which is forested, 7% arable land and 17.7% grassland. It is also a region of distinctive subregions characterized by similarities in their geographic, ethnic, developmental or political characteristics. The major subregions are Mexico and Central America, the Andean countries, the southern Cone and the Caribbean. Table 10.1 gives selected economic and human welfare indicators for these subregions.

The following ecoregions can be identified: the Tropics, the Andes, the southern Cone and the arid and semi-arid areas.

Approximately 77% of Latin America and the Caribbean (LAC) is tropical. The tropical ecoregion can be subdivided into three major agro-ecosystems: the humid tropics, the subhumid tropics and the intermediate valleys and plateaus. The humid tropics include two areas: the humid lowlands, which are less than 600 m above sea level (m a.s.l.); and the humid hillsides, which are between 600 and 1200 m a.s.l. The subhumid zone also includes lowlands and hillsides. The main difference between the humid and subhumid zones is in total rainfall and its annual distribution. Long dry periods of up to 6 months are typical in the subhumid tropics. The intermediate valleys and plateaus are located at 1200–2800 m a.s.l. and are characterized by lower mean temperatures than the humid and subhumid tropics. The tropical areas contain most of the forest areas of the region and hence population density is low. Livestock production is an important activity. Beef cattle are usually produced in extensive grazing systems. Milk

Table 10.1. Selected economic and human welfare indicators according to subregions in Latin America and Caribbean (LAC).

	Mexico and Central America	Andean zone	Southern Cone	Caribbean
Gross national product (US$ 10[6])	200,315	123,855	479,204	33,542
Income per capita (US$/year)	1,436	1,158	2,132	1,216
Urban population[a]	52.3–90	65.5–78.6	74.4–81.0	54.3–70.6
Poverty incidence	63.4	53.6	37.4	81.2
Human development index (HDI)[b]	0.67	0.66	0.81	0.64
Agricultural land distribution[c]	0.80	0.74	0.86	0.75
Calorie consumption per capita (calories day^{-1})	2,548	2,369	2,784	2,631
Protein consumption per capita (g day^{-1})	65.4	57.9	79.8	73.8

Source: adapted from Winograd (1995).
[a]1990 and its projection to 2030.
[b]HDI; life expectancy, education and health.
[c]Gini coefficient.

is produced mainly in dual-purpose (dairy–beef) production systems and it is usually a smallholder activity.

The Andean ecoregion covers around 2 million km^2 in western South America, from Venezuela and Colombia in the north to Chile and Argentina in the south. It is a mountainous area characterized by very steep slopes and extreme climatic variations due to the interplay of both latitude and altitude.

This ecoregion contains the poorest population of LAC. Agricultural production and productivity are constrained by the size of farmers' land holdings, climate, steep slopes and inadequate infrastructure, which hampers marketing. Beef is produced mainly in smallholder mixed systems. Milk is produced in intensive systems in the inter-Andean valleys and savannas, near major urban centres. Sheep and South American camelids have comparative advantages over cattle at higher altitudes.

The southern Cone is a fairly extensive area, going south from subtropical zones to mostly temperate areas in south-eastern South America. It includes southern Brazil, Paraguay, Uruguay, Argentina and southern Chile, and is the wealthiest agricultural ecoregion in LAC. Farm size is usually not a constraint. Soils are good and the climate is mostly temperate. These favour agricultural systems of high productivity with relatively low inputs. The lack of steep slopes allows mechanization. This ecoregion has the highest concentration of beef cattle and sheep in LAC, with most being kept in extensive production systems. Specialized milk production systems based upon grazing are also common.

The arid and semi-arid areas of LAC are mainly in northern Mexico, north-eastern Brazil and in the coastal regions of Peru and Chile. Where there is no access to irrigation, poverty prevails despite the population density being lower than in the Andes. Goats are the main livestock in non-irrigated areas. They are produced mainly in either transhumant or nomadic herds. Intensive milk production is practised in irrigated areas, especially those close to urban centres.

The human population of LAC was 466 million in 1993 (FAO, 1994) and it is expected to reach 700 million in the year 2025. The rural population is expected to remain at the current level, all the increase being in the urban population (Fig. 10.1). Therefore, the pressure on rural areas to produce more food will increase. Food demands will not be satisfied unless there is a drastic change in natural resource use, through more appropriate policies and technologies. Changes are needed to improve infant and adult nutrition, health and sanitary conditions and education.

This situation is particularly true in the Andes and the tropical hillsides and lowlands. Although total food production in the region has increased in the past 20 years, food production per capita has remained constant (FAO, 1996). The vast majority of people in the region have calorie and protein intakes well below recommended levels. In the 1980s, about 55 million people in LAC suffered from malnutrition (Janssen, 1991). Table 10.1 shows the average per capita consumption of calories and protein in LAC (FAO, 1994) and the difference in economic and other welfare indicators among

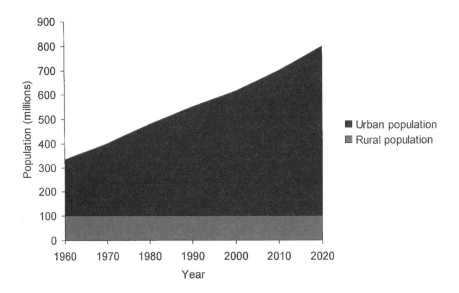

Fig. 10.1. Evolution of rural and urban population in Latin America and the Caribbean. (World Bank, 1992).

subregions. However, this table which already shows deficits, fails to show the differences in consumption linked to disparities in income. As a typical example, Fig. 10.2 (Ruiz, 1995) shows milk consumption according to socio-economic strata in Guatemala. Over 65% of the population have an annual income of less than US$2400 and consume less than 5 litres of milk per capita per year, while less than 2% of the population have an annual income of more than US$12,000 and consume approximately 90 litres of milk per capita per year.

Poverty is increasing in the region; hence malnutrition can be expected to increase. The increase in malnutrition will be aggravated by the continuing degradation of natural resources used for agriculture and consequent declining agricultural yields. For instance, soil erosion in the Peruvian Andes is proceeding at a rate well above the 5 t ha^{-1} threshold value (Fig. 10.3) (Quiroz *et al.*, 1995). This indicates that several of the current management practices are not sustainable. An exception is the use of rangelands for livestock production, which has the lowest erosion rates.

THE ROLE OF LIVESTOCK IN DEVELOPMENT

LAC is home to large numbers of livestock (Table 10.2). Over 25% of the cattle population of the world is found in LAC, the majority being in the southern Cone, followed by the Andean region. Over 18% of the world's dairy cows are found in LAC.

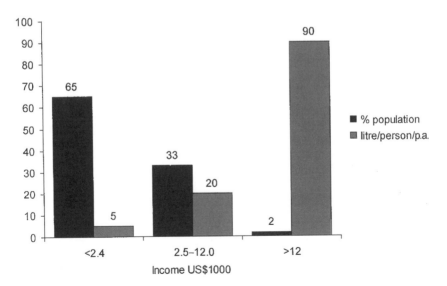

Fig. 10.2. Milk consumption (litre per person per annum) by socio-economic strata, Guatemala, 1981. (Ruiz, 1995).

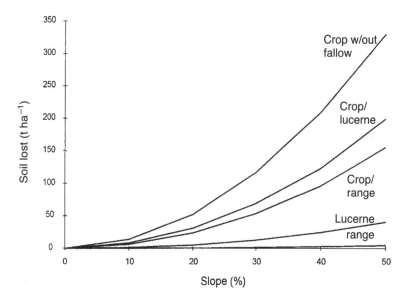

Fig. 10.3. Estimates of soil losses under different land-use systems in the highlands of Peru. (Adapted from Quiroz *et al.*, 1995.)

Table 10.2. Resources in animal agriculture in Latin America and the Caribbean (LAC) by country groups.

	World	Mexico and Central America	Caribbean	Andean	Mercosur[a]	LAC (%)
Cattle (10³ head)	1,277,793	41,124	9,086	58,110	221,837	25.8
Milking cows (10³ head)	223,634	8,223	1,018	8,013	23,804	18.4
Goats (10³ head)	591,802	11,143	2,438	6,758	16,002	6.1
Sheep (10³ head)	1,110,782	5,746	1,057	28,722	70,274	9.5
Pigs (10³ head)	870,705	19,370	3,271	13,235	36,388	8.3
Chickens (10⁶ head)	11,868	345	119	332	699	12.6
Permanent pastures (10³ ha)	3,424,323	88,487	7,731	130,553	364,020	17.3
Forested and woodlands (10³ ha)	3,879,796	56,486	42,522	221,100	560,519	22.7

Source: FAO (1994).
[a]Mercosur: includes Brazil, Argentina, Uruguay and Paraguay.

LAC contains over 17% of the world's grasslands and 22% of its forest and woodlands. It is therefore not surprising that beef exports are important, especially for the southern Cone and Central America. In 1993, LAC exported over 1.5 million tonnes (Mt) of beef, generating an income of US$2.95 billion (FAO, 1993). In contrast, the region is a net importer of milk and dairy products, in 1993 importing 0.8 Mt of milk, 82,395 t of butter and 87,352 t of cheese, at a cost of US$1.56 billion (FAO, 1993).

Despite the importance of livestock in LAC, their productivity is well below that obtained in developed countries (Table 10.3). Technological research has shown that there is still significant potential for improvement.

The greatest need for livestock development efforts to alleviate poverty is in the Andean ecoregion and the tropical forest margins. Hence, this chapter concentrates on the problems and challenges in these two areas.

Results from Puno, Peru, on the high plateau of the Andes, indicate that livestock are the major source of income for peasant families (Table 10.4). Livestock traditionally have also protected peasant economies from climatic and economic uncertainty by providing a means of storing surpluses from good seasons for use in times of need. Manure and animal traction are critical inputs to crop production. Income from livestock can be invested in crop production and funds education for peasant families. Women peasants play a critical role in livestock production. Investments in livestock research and development can thus have a very significant effect on rural development.

In the tropical areas, livestock help stabilize agricultural systems. Smallholders in the tropical zone commonly practise shifting cultivation. After they have extracted precious woods from the forest, they burn the remaining trees. This mineralizes the organic matter and enriches the soil, resulting in high yields of annual crops (maize, rice) for a couple of years. After that, soil fertility declines sharply and the smallholders plant pastures to support beef, or beef and milk production. As access to markets is often difficult,

Table 10.3. Yield of livestock products in the American continent (kg head^{-1} year^{-1}).[a]

Region/country	Pork	Lamb/mutton	Beef	Cow milk
Canada and USA	126	14.4	102	6564
LAC	37	2.6	33	1095
Argentina	49	3.2	52	2559
Costa Rica	45	—	45	1365
Ecuador	16	4.4	25	1700

[a]Estimates are based on FAO (1990) data. The values were obtained by dividing the total reported production by the total population of each species, disregarding animal category; thus, the actual yield values must be higher for every case.

Table 10.4. Sources of income and expenditure for a peasant family in the community of Santa Maria, Ilave, Puno, Peru, 1992.

Subsystem	US$ year^{-1}	%
Income		
Crops (potato, quinoa, oca, barley, others)	214	21.4
Animal production	458	45.8
Processing (handicraft, animal products)	107	10.7
Migration and trading	60	6.0
External support (food aid, others)	162	16.1
Total gross income	1001	
Expenditure		
Self-consumption of products	393	44.4
Food and supplies	110	12.4
External support (food aid and others)	162	18.3
Other cash expenses	220	24.9
Total gross expenses	885	
Gross margin	116	

Adapted from PISA (1993) and PRODASA (1994).

smallholders generally prefer beef production to dairying. However, small-holder dairying is becoming increasingly important, especially in peri-urban areas. In tropical Latin America, milk is produced mainly in dual-purpose (dairy–beef) cattle production systems. Dual-purpose cattle comprise over 78% of the cattle in tropical areas and produce over 41% of the milk.

The above-described situation and the examples provided below show the interlinkage of poverty, livestock and ethics. Livestock are critical to the sustained livelihood of peasants, for whom they are the main source of income and savings. Livestock products are in high demand by an ever growing urban population.

Truly participatory development efforts must consider farmers' needs and aspirations, as well as consumer demand. It is unethical (and is likely to be unproductive) to impose development goals that do not consider the circumstances of stakeholders and their own goals.

Many policy measures enacted in LAC have created an unfavourable environment for smallholder agriculture and have contributed to the vicious cycle of deterioration of natural resources, low agricultural productivity, poverty, inequity, migration, social deterioration and violence (Fig. 10.4).

Ill-advised policies

Examples are now given of policies which have been followed in the past with poor or negative results and are now considered to be ill-advised and unethical policies.

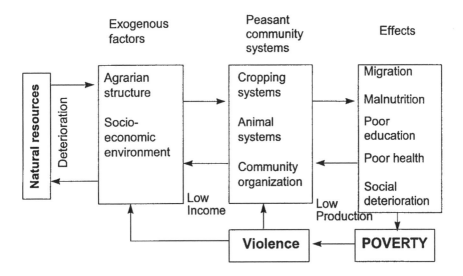

Fig. 10.4. The vicious cycle of deterioration of natural resources, low agricultural productivity, poverty and violence. (Adapted from PISA (1993) and PRODASA (1944)).

Dumping
Beef and milk production have been highly subsidized in developed countries, and beef and milk in excess of demand has been 'dumped' in developing countries, posing unfair competition for local producers. As a result, smallholders could not afford to invest to intensify their production systems. Livestock have thus been confined mainly to savings and maintenance roles in many developing countries.

Land reform
In Peru, land reform was supposed to improve equity among sectors of society by promoting redistribution of wealth. However, it acted against livestock development by causing a dramatic decrease in the livestock population (Leon-Velarde and Quiroz, 1994). It neither improved access to land for poorer sectors of the population nor improved livestock productivity. Other attempts at reforming land tenure on the continent also failed to promote equity, as the poorest often did not gain access to more land, or the land they did acquire was unproductive.

Price controls
Price controls have been implemented to protect the consumer. However, not enough consideration was given to their effect on producers. By keeping prices low, price controls favoured consumers, especially in the cities, while smallholder farmers in effect subsidized intermediaries in the marketing chain and consumers. Artificially low product prices discouraged intensification of livestock production systems.

Urban-oriented development policies

Most countries in the region have encouraged the expansion of urban centres by concentrating investments and services (roads, electricity, potable water, sanitation, health and education) in the cities. As the cities grew and became critical constituencies, many governments tended to continue subsidizing or investing in the urban areas, promoting further migrations from rural areas to urban centres. The lack of infrastructure in rural areas has hampered marketing of agricultural products and discouraged investments in the more marginal areas and in the livestock sector.

Rural development policies

It can be argued that, in most cases, rural development policies did not work. Policies in general have been established with little participation of stakeholders. Results have not favoured society, nor have they helped to achieve development goals. Examples of such policies include the establishment of subsidized credit for agriculture at times of high inflation, which resulted in farmers investing in more profitable enterprises such as trading and speculating with hard currency. Another example is the formulation of policies in the Amazon to expand the agricultural frontier. This resulted in large investments in infrastructure, which encouraged large-scale ranchers to clear large tracts of forest and to speculate on land. This in turn has created a bad reputation for livestock development. It is estimated that over 30,000 km^2 of forests in Brazil were destroyed each year throughout the 1970s. After these policies were removed, the rate of deforestation fell to approximately 10,000 km^2 per year (Niamir-Fuller, 1997).

Conflicting policy objectives and shifting policies

The lack of consistency of policies has often created instability and an unfavourable environment for livestock development. As a result, smallholders have preferred to avoid intensification and its inherent risks. Fiscal and monetary policies and protection of industries have contributed to the poor performance of agriculture, particularly the livestock sector.

Lack of coordination in trade policies across subregions

Although different subregional agreements have been established to harmonize trade and economic policies, lack of coordination among them has been common. This has resulted in unfavourable and even illegal trade across borders, with some countries trying to subsidize food while others followed a more liberal pricing policy. Many governments are realizing these mistakes. There is now a clear trend towards more liberal economic policies and strengthening regional trade blocks such as NAFTA (North American Free Trade, which involves Canada, the USA, Mexico and Chile), SIECA (Central American Economic System), the Andean Pact and MERCOSUR (which involves the southern Cone countries).

Research and extension policies

Most countries of the region recognize the importance of supporting re-search and extension efforts to search for solutions to the problems of smallholders. However, most of them have promoted discipline-oriented or commodity research. These approaches have not proved effective in addressing the complex problems of smallholders. Multidisciplinary research is needed to address such problems, where socio-economic and environmental problems may be closely intertwined. To make matters worse, structural adjustment programmes are reducing the size of the government sector in the short term. While the private sector is taking care of the needs of intensive agricultural systems, it plays a very minor role in supporting smallholder agriculture, especially in marginal areas. Therefore, research and development efforts for the smallholder sector have decreased substantially, especially during the last 5 years. The level of international aid has also decreased during that period, as other pressing issues appear in the agenda of donor countries, including environment, human rights and participation of civil society, poverty alleviation, trade and development. Most institutions have failed to adapt to the changing environment, thus failing to obtain internal resources to support research and extension.

Macro–micro linkages

Often, there is lack of coherence between the formulation of macro-economic policies (e.g. free trade, exchange rate control) and the formulation of micro-economic policies (e.g. smallholder farm development). Macro-economic policies often have a negative impact at the farm level, resulting in contradictory policy goals. For example, efforts to minimize the budget deficit may inadvertently reduce marketing of livestock products because of reducing public expenditure on infrastructure.

Environmental policies in the developed countries

Farmers in developed countries commonly employ very intensive livestock production systems. These systems have often resulted in soil, water and air pollution, especially in peri-urban areas; hence, strict environmental regulations have been introduced to control such systems. Also, environmental groups from developed countries campaign against livestock and express strong concerns about livestock development efforts in developing countries. Generalizations are made based on few empirical data, such as the destruction of the Amazon forest to promote cattle ranching or the role of goats in overgrazing of grasses and shrubs in semi-arid regions. Those concerns give livestock development a bad image, reducing investment and aid to developing countries. They do not consider the complex interactions of human needs, livestock and the environment. They ignore the damage that is done to smallholder development efforts in other areas where the problems are most acute and the fact that livestock play a critical role in ensuring the sustainability of those systems.

Nutritional considerations

People in the developed world tend to consume too many livestock products and demand is stagnant (Steinfeld *et al.*, 1997), whereas most people in developing countries still consume too few livestock products for a healthy diet and demand for livestock products is still unsatisfied. However, environmental and nutritional concerns, mainly in developed countries and followed by small but vocal groups in developing countries, put pressure on governments to curtail livestock development efforts in developing countries.

Public sector issues

Infrastructure and institutional deficiencies are much more severe in rural areas than in urban areas; social services are fewer in rural areas than urban areas. Investment in market infrastructure reduces costs and risks across a broad range of outputs and inputs, and hence would have extensive effects on rural development.

THE ROLE OF RESEARCH TO ADDRESS POVERTY ALLEVIATION, EQUITY AND PROTECTION OF THE NATURAL RESOURCE BASE

The previous paragraphs show how research and development efforts have often failed to find sustainable solutions to the problems of poverty and equity, the role of livestock to address them, and the protection of the natural resource base. It is clear that development efforts have to be knowledge based. It is also clear that, in order to be of benefit, knowledge has to be generated through holistic, multidisciplinary and participatory approaches. A thorough understanding of the causes of the problems and their complex nature is essential if viable and ethical solutions are to be proposed.

Over the last 20 years, systems research approaches have been promoted in LAC in an effort to find solutions to the problems facing smallholder agriculture. For example, the International Development Research Centre (IDRC) of Canada played a critical role in funding and promoting the Latin American Animal Production Systems Network (RISPAL). The network involves 15 projects in 11 countries of the region and is coordinated by the Inter-American Institute for Cooperation in Agriculture (IICA). Projects followed a common approach that included multidisciplinary and participatory research linked to development activities. Projects implemented the following steps: diagnosis of problems and opportunities; design of potential interventions (including ex-ante assessment of potential bioeconomic and social impacts); on-farm testing of best bets; and transfer of technologies that had proved successful under farm conditions. These experiences have been reported by Li-Pun and Sere (1993) and Ruiz (1993).

Lessons learned from systems research approaches

Some important lessons have been learned from applications of systems research, which are now summarized here.

- Systems research is an important tool for gaining a better under-standing of the problems and opportunities facing the smallholder farmer. However, it is not a panacea; problems are quite complex and there are no quick fixes.
- Systems research was conducted mostly at the farm level. Often, however, the problems require solutions off-farm, such as addressing marketing problems or policy issues.
- Promising research results have been tested on farms. Substantial bio-economical gains have been demonstrated, especially on small- and medium-scale farms.
- Technological solutions alone will not solve the problems of small-holder agriculture in marginal areas; both policy and technological solutions are required.
- Interinstitutional linkages are becoming increasingly important. Many research and development projects attempted to link research and development institutions. Some were successful, many not. Participation in those efforts was mostly from the government sector. The increasingly important role that non-governmental organizations (NGOs) are playing in development, and in some cases in research, was not always considered appropriately.
- Research and development efforts have to be sustained over a long time. Systems research methods were being developed and availability of trained staff was scarce. Much had to be learned on the job and through short-term training.
- Institutionalization of systems approaches was particularly difficult. As was mentioned, most institutions in LAC follow disciplinary oriented and commodity-oriented approaches. However, there are now many systems research practitioners in the region, as a result of RISPAL and other efforts such as those of RIMISP (The International Network on Methodology for Production Systems Research), also supported by IDRC, bilateral projects, the USAID Small Ruminant Collaborative Program, the collaborative work supported by CIRAD and ORSTOM, etc.

Organization of ecoregional research

In 1991, the Technical Advisory Committee of the Consultative Group on International Agricultural Research (CGIAR) proposed the use of eco-regional research to tackle the problems of agricultural productivity and sustainable use of natural resources simultaneously. Ecoregional research was defined as holistic and multidisciplinary research to address problems of ecosystems that are geographically bound.

In March 1992, the International Potato Center (CIP) organized a meeting to review experiences and perspectives for the development of the Andean ecoregion. Over 70 participants representing 35 institutions attended the meeting. The participants recommended the creation of a consortium of institutions to address problems of natural resource management in the Andean ecoregion, given its paramount importance to the livelihood of the people in the highlands and its effects on adjacent ecosystems (including the tropical areas, to which many impoverished peasants from the highlands migrate). The consortium, CONDESAN (Consortium for Sustainable Development of the Andean Ecoregion), originally was organized and supported by CIP, IDRC and the Swiss Agency for Development and Co-operation (SDC). Suggestions for its organization were proposed by Li Pun and Paladines (1993). Several other donors and international organizations joined later, including Denmark, Germany, the Inter-American Development Bank, The Netherlands, Spain, the United States Agency for International Development–Collaborative Research Support Program (USAID–CRSP) and others. At present, over 90 institutions are members of the consortium.

CONDESAN operates in five benchmark sites representative of the Andean ecoregion in Venezuela, Colombia, Ecuador, Peru and Bolivia. The main themes addressed are land and water management, biodiversity (Andean roots and tubers, pastures), production systems (crop–livestock) and policies. An electronic information system (INFOANDINA) links partners.

A small grant scheme provides matching funds to leverage resources from participating institutions and encourages interinstitutional linkages. Participating institutions include national agricultural research institutes (NARIs), universities, NGOs, bilateral and multilateral agencies and international organizations.

CONDESAN works at different system levels: region, watershed (benchmark sites) and farm (Fig. 10.5). The consortium uses modelling and geographic information systems to characterize production systems and to make ex-ante assessments of potential biophysical and socio-economic impacts, and employs participatory diagnosis techniques, on-farm testing and local round-tables to encourage stakeholders to participate in identifying problems, opportunities and concerted actions. It also links research, policy formulation and development actions. The consortium's actions are guided by the following principles: openness of participation, entrepreneurship (sharing of costs and opportunities), efficiency (avoiding duplications and building from existing experiences) and transparency (participatory management).

Within this consortium, the International Livestock Research Institute (ILRI) collaborates in the livestock activities. Activities aim to improve

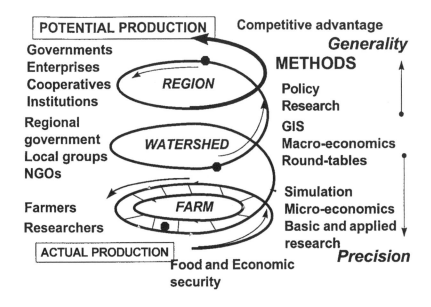

Fig. 10.5. System hierarchies and research methods utilized by the Consortium for Sustainable Andean Development, CONDESAN. (Leon-Velarde and Quiroz, 1997).

crop–livestock systems in which milk is the main livestock product. Milk is the product with highest comparative advantages over other agricultural commodities in most of the Andean ecoregion. In the high-altitude zones, e.g. the Altiplano of Peru and Bolivia, South American camelids (alpaca, llama) have the comparative advantage.

Results from the CONDESAN programme

Preliminary results from CONDESAN are now available, and some indications of progress can be identified, as follows.

- The consortium is creating awareness of the need for appropriate natural resource management in the highlands and the effects of natural resource management on erosion, water and energy balance. The effects of poverty in the highlands exacerbate the negative effects of migration and social and environmental deterioration in adjacent ecosystems (tropical lowlands as well as coastal areas in the case of Peru). Sustainable development of the Andean region will not be possible if problems of poverty and inequity persist.
- The consortium is a novel and viable model for interinstitutional collaboration to address complex problems of increasing agricultural productivity, sustainability and equity.

- Computerized models developed allow the integration of information and estimate the impacts of improving agricultural productivity on the above-mentioned goals while ensuring sustainability and equity at different hierarchical levels.
- Ex-ante assessments show that in some marginal areas the only possibility for substantially increasing smallholder income is through changes in policies (e.g. water pricing in the cities and provision of subsidies to smallholders in the highlands) and improvement of markets. Those policy changes would favour the adoption of technologies.
- CONDESAN has, with relatively limited resources, leveraged substantial resources from partners, thus allowing holistic and concerted actions addressing problems in the Andean ecoregion.

Much remains to be done. However, this novel initiative has shown promise in linking research and development actions, developing policy and technology interventions, encouraging participatory research and development and promoting wide interinstitutional collaboration.

ETHICS AND QUALITY OF LIFE

Ethics becomes a major issue when some groups within society benefit while others do not. Throughout this chapter, it has been argued that in spite of its wealth in natural resources, LAC does not realize its potential, and significant problems exist in terms of poverty, quality of life and equity.

It is evident that in order to change this situation, major alterations are required in the way in which society views problems, their causes and solutions. Some sectors of society would argue that the causes are socio-cultural, and that the long-term solution is through education; others, that the solution is economic and that by improving the industrial sector in cities, wealth and employment opportunities would be developed for the people that migrate from the rural to urban areas. This hypothesis is backed up by examples from industrialized societies. However, the costs associated with the displacement of people from rural to urban areas, and the difficulties that people face due to living in a strange environment (the 'social costs') are seldom addressed. The analysis also fails to address urban–rural linkages, and the inequalities generated by present development policies.

These issues need to be understood better by society. Education on these issues and the search for solutions is required. These may guide governments, development agents and the general public in the search for solutions that promote equity and equal opportunities within society.

This type of education can be promoted at least at two levels. Within the university curricula, development courses should be incorporated independently of the area of specialization. Those courses need to take a more holistic approach that incorporates the ethics of development, along with the socio-economic approaches for development of rural and urban societies, and their interdependence. Global interdependence also needs to be addressed, for example in analysing the effects of international trading policies on the production sector in developing countries.

Public awareness campaigns should be organized to educate the general public on issues of ethics in development and quality of life for both the urban and rural populations. Cases such as the effects of policies mentioned in this chapter could be documented and communicated to the general public, so mistakes would not be repeated. The same could be said for the understanding of rural–urban linkages, in societies that have followed policies that encouraged centralization and the development of large urban centres to the detriment of the quality of life for large sectors of society.

Failure to link ethical issues with development and quality of life would offer only partial solutions to problems, and may lead to very detrimental effects in the long run, and unsustainable livelihoods.

CONCLUSIONS

Latin America is a highly diverse region. In consequence, generalizations about its problems and opportunities are not appropriate. Recognition of subregions is critical in addressing problems.

In spite of the region's wealth in terms of natural resources for livestock production, poverty in the rural areas is still acute, especially in the Andes, tropical forest margins and semi-arid ecoregions. Livestock are an important asset in both macro- and micro-economic terms, and can contribute substantially to improving the quality of life of rural people.

Past policies have harmed efforts to promote equity and sustainable development of the livestock sector. There are a number of changes in the behaviour of different sectors of society that could help to improve the role that livestock play in the livelihood of smallholder farmers and contribute to alleviating poverty and environmental degradation and improving equity. Some of these changes are discussed below.

Rural investments

Investments in rural infrastructure (roads, electricity, communications, education, credit and extension) can promote rural development by facilitating marketing of agricultural inputs and products and development of agro-industries and by providing access to relevant information which is essential to the population. Processing agricultural products, especially livestock products, can provide incentives for livestock development and contribute to rural development and equity.

Tax reforms and environmental policy

Appropriate tax schemes need to be developed to ensure that ranchers, large-scale farmers and the urban population contribute to the development of the infrastructure of benefit to smallholders in marginal areas. For example, appropriate water pricing in the cities could generate funds for investment in infrastructure in marginal areas in the highlands. Without those investments, the vicious cycle of poverty and environmental deterioration cannot be broken. Assessing appropriate taxes to ranchers could stimulate a more efficient use of resources, as well as the sale of land to peasants.

International and national investments

The need for international investments still remains, especially for research and development activities targeted towards smallholder agriculture. This includes the development of infrastructure, credit, research and extension, etc. However, much can also be achieved by promoting national savings and private investment. The former may require modification of tax policies, to increase government revenues through taxes, and the establishment of an enabling environment for private sector investments.

International trade

Dumping of cheap food to developing countries should be stopped as it hampers agricultural development in the recipient countries.

New approaches towards interinstitutional cooperation in research and development

The role of research for development is now better understood. Research to address the complex problems of smallholders needs to be holistic, multidisciplinary and participatory. Past lessons from systems research are being applied by an ecoregional consortium of institutions, CONDESAN, that is attempting to increase agricultural productivity while promoting sustainable management of natural resources and equity in the Andes. Preliminary results from CONDESAN indicate that it is a viable mechanism for interinstitutional collaboration involving a wide range of stakeholders interested in the development of the Andean ecoregion. This interinstitutional cooperation has to be based on openness, sharing of costs and benefits, use of institutional comparative advantages and transparency in operations.

Environmental and nutritional concerns in developed countries

Perceptions in developed countries of the negative environmental and nutritional effects of livestock have reduced investment in livestock development in developing countries. This hampers efforts to alleviate poverty and promote equity, especially in areas where the needs are greatest.

Developments that are likely to facilitate livestock development include: formation of subregional trading blocks, application of more liberal economic policies, articulation of macro- and micro-economic policies, increased access to information and education and trends towards democratization. Within these major changes, opportunities will arise to promote poverty alleviation, equity and environmental protection. Many of the negative trends could be overcome, leading to a much better and sustainable livelihood in the region.

SUMMARY

Latin America is well endowed with natural resources, yet poverty is widespread, wealth is distributed unequally and natural resources in marginal areas are deteriorating. The region can be broken down into several subregions according to geographic and ethnic characteristics, relative level of human development and/or political organization. These differences must be recognized if efforts to solve common problems are to be targeted effectively.

In spite of the region's relative wealth, nutritional deficits, especially in the consumption of calories and animal protein, are widespread. Problems of poverty, environmental degradation and equity are most acute in the Andean ecoregion and the tropical forest margins, especially in the rural areas. Livestock play a critical role in providing both food and non-food products for an ever-increasing population, and in ensuring the livelihood of impoverished smallholder farmers. Development of the livestock sector could help improve the economic and nutritional status of the poorest sectors of the population and promote equity and environmental protection.

Many policies applied in the past have worked against these development goals. This chapter provides examples of international and national policies that have had negative effects on livestock and rural development and suggests ways in which policies and institutional arrangements might be improved. Development problems will be solved only through the application of better policies, the use of appropriate technologies and interinstitutional cooperation. Achieving this requires implementation of holistic and participatory approaches that link research and development efforts.

ACKNOWLEDGEMENTS

The authors wish to acknowledge the collaboration of Dr Simeon Ehui of ILRI and Dr Elias Mujica of CONDESAN in reviewing this manuscript, and Mr Paul Neate of ILRI for assistance in editing.

REFERENCES

Food and Agriculture Organization of the United Nations (1990) *Yearbook of Production*, Vol. 44. FAO, Rome, Italy.

Food and Agriculture Organization of the United Nations (1993) *Yearbook of Trade*, Vol. 47. FAO, Rome, Italy.

Food and Agriculture Organization of the United Nations (1994) *La Política Agrícola en el Nuevo Estilo de Desarrollo Latinoamericano*. Oficina Regional de la FAO para America Latina y el Caribe, Santiago, Chile.

Food and Agriculture Organization of the United Nations (1996) *Food Balance Sheets*. FAO, Rome, Italy.

Janssen, W. (1991) Economic trends in Latin America and the Caribbean: implications for agriculture and the generation of agricultural technology. In: *CIAT in the 1990s and Beyond: a Strategic Plan* (Supplement). CIAT (Centro Internacional de Agricultura Tropical), Cali, Colombia, pp. 1–13.

Leon-Velarde, C. and Quiroz, R. (1994) *Análisis de Sistemas Agropecuarios: Uso de Métodos Bio-matemáticos*. Centro de Investigación de Recursos Naturales y Medio Ambiente, La Paz, Bolivia.

Leon-Velarde, C. and Quiroz, R. (1997) Estrategia operacional en el marco conceptual del CONDESAN en la Ecorregión Andina. CONDESAN/CIP, Lima, Peru.

Li-Pun, H.H. and Paladines, O. (1993) El rol de las pasturas y la ganaderia en la sostenibilidad de los sistemas de produccion andina. In: *Centro Internacional de la Papa. El Agroecosistema Andino: Problemas, Limitaciones, Perspectivas*. Anales del Taller Internacional sobre el Agroecosistema Andino, CIP, Lima, March 30–April 2, 1992, pp. 187–212.

Li-Pun, H. and Sere, C. (1993) Animal production systems research in developing countries: overview and perspectives. In: *Proceedings of the VII World Conference on Animal Production*. University of Alberta, Edmonton, Canada; pp. 329–348.

Niamir-Fuller (1997) Case study presented at the *Electronic Conference on Livestock, Environment and Human Needs*. ILRI-FAO-INFORUM-IICA and the World Bank, Addis Ababa, Ethiopia.

PISA (1993) *Proyecto de Investigación de Sistemas Agropecuarios Andinos*. Informe Final. Convenio (INIAA-CIID-ACDI) Instituto Nacional de Investigaciones Agricola y Agroindustrial; Centro Internacional de Investigaciones para el Desarrollo; Agencia Canadiense para el Desarrollo Internacional, Lima, Peru.

PRODASA (1994) *Proyecto de Desarrollo Agropecuario Sostenible en el Altiplano*. Convenio (CIP-CIRNMA-CIID) Centro Internacional de la Papa; Centro de Investigaciones en Recursos naturales y Medio Ambiente; Centro Internacional de Investigaciones para el Desarrollo, Lima, Peru.

Quiroz, R., Estrada, R.D., Leon-Velarde, C.U. and Zandstra, H.G. (1995) Facing the challenge of the Andean Zone: the role of modeling in developing sustainable management of natural resources. In: Bouma, J., Kuyvenhoven, A., Bouman, B.A.M., Luyten, J.C. and Zandstra, H.G. (eds), *Ecoregional Approaches for Sustainable Land Use and Food Production.* Proceedings of a Symposium on Eco-regional Approaches in Agricultural Research, 12–16 December 1994, ISNAR, The Hague. Kluwer Academic Publishers in cooperation with International Potato Centre, Doldrecht, The Netherlands, pp. 13–31.

Ruiz, M.E. (1993) Animal production systems experiences in Latin America. In: *Proceedings of the VII World Conference on Animal Production.* University of Alberta, Edmonton, Canada, pp. 387–411.

Ruiz, M.E. (1995) The Latin American livestock sector and research prospects. In: Gardiner, P. and Devendra, C. (eds), *Global Agenda for Livestock Research. Proceedings of a Consultation.* Nairobi, Kenya, 18–20 January, 1995. ILRI (International Livestock Research Institute), Nairobi, Kenya, pp. 61–73.

Steinfeld, H., de Haan, C. and Blackburn, H. (1997) *Livestock–Environment Interactions: Issues and Options.* European Commission, Directorate-General for Development Produced by WREN media, Fressingfield, UK.

Winograd, M. (1995) Indicadores Ambientales para Latinoamerica y el Caribe: Hacia la Sustentabilidad en el Uso de Tierras. *En colaboracion con: Proyecto IICA/GTZ, Organizacion de los Estados Americanos,* Instituto de Recursos Mundiales, San Jose, C.R.

World Bank (1992) *World Development Report 1992. Development and the Environment.* Oxford University Press, Oxford.

Livestock, Ethics, Quality of Life and Development in Africa

11

George K. Kinoti[*]

Department of Zoology, University of Nairobi, Nairobi, Kenya

INTRODUCTION

If our scientific work is to have real significance, it must go beyond publishing a paper or obtaining the next research grant. It must contribute to human well-being. I believe that we have a particular responsibility to ensure that our work contributes to the improvement of the life of the poor and disadvantaged. We all need time to stop and think about the contribution of our work to the quality of life, particularly in poor countries.

In this chapter, we will first look briefly at the quality of life in Africa today. Secondly, we will consider the contribution of agriculture, including animal agriculture, to African welfare and development. Finally, we will consider some issues related to livestock production in Africa. I should explain here that this chapter is concerned only with Africa south of the Sahara, or black Africa as it is sometimes called.

QUALITY OF LIFE IN AFRICA TODAY

Quality of life and ethics

This chapter deals primarily with the quality of social and economic life in Africa. However, we must bear in mind the fact that human life is a complex of economic, social, political, cultural, moral, intellectual and spiritual components. Each of these aspects of life affects, and is in turn affected by, all the other aspects.

*Present address: African Institute for Scientific Research and Development (AISRED), PO Box 14663, Nairobi, Kenya.

For example, moral values and ethics (i.e. conduct issuing from moral values) strongly influence or even determine socio-economic life. Thus, although the world has enough resources for everyone to enjoy a decent standard of life, at least one out of every four human beings lives in absolute poverty due largely to inequitable distribution both within and among nations. The eminent American economist, John Kenneth Galbraith (1998) states moral responsibility for poverty with great clarity. He says,

> The problem is not economics; it goes back to a far deeper part of human nature.
>
> As people become fortunate in their personal well-being, and as countries become similarly fortunate, there is a common tendency to ignore the poor. Or to develop some rationalization for the good fortune of the fortunate. Responsibility is assigned to the poor themselves. Given their personal disposition and moral tone, they are meant to be poor. Poverty is both inevitable and deserved. The fortunate individuals and fortunate countries enjoy their well-being without the burden of conscience, without a troublesome sense of responsibility. This is something I did not recognize writing [the famous book *The Affluent Society*] 40 years ago; it is a habit of mind to which I would now attribute major responsibility.

In 1997, the United Nations Development Programme (UNDP) stated, 'Although poverty has been dramatically reduced in many parts of the world, a quarter of the world's people remain in severe poverty. In a global economy of $25 trillion, this is a scandal – reflecting shameful inequalities and inexcusable failures of national and international policy' (UNDP, 1997). One may add that the inequalities and failures are to a large degree attributable to economic or political interests at the personal, corporate or national level.

The widespread poverty in the so-called developing countries is a major moral issue. It is to a large extent due to unethical practices on the part of political leaders, business leaders and public servants in both the developed and the developing countries. George (1986) in her book *How the Other Half Dies: the Real Reasons for World Hunger*, discusses the selfish political and economic reasons for hunger in a world where there is surplus food. Widespread misuse of the so-called aid money is discussed by Hancock (1991) in his book, *Lords of Poverty*.

Although this chapter is concerned with Africa, we should remember that poverty is a global problem. Some 1.3 billion people in the developing world, or about one-third of the population, live on *less than US$1 a day*. Of these, over 950 million live in Asia, 220 million in Africa south of the Sahara, and 110 million in Latin America and the Caribbean. In addition, over 100 million people in industrial countries and 120 million in Eastern Europe and the Commonwealth of Independent States live below the poverty lines set for those countries. Some 37 million people in industrial countries are jobless (UNDP, 1997).

Socio-economic conditions in Africa

Africa is endowed with sufficient natural resources – land, minerals, forests, water resources, wildlife, etc. – for the people to be able to enjoy a good standard of life. Yet Africa is oppressed by widespread poverty, hunger, disease and other problems of socio-economic underdevelopment. The depressing socio-economic conditions are clearly indicated by the gross national product, gross domestic product, health status, education status and the human development index.

Gross national product
Nearly 80% of the African countries are classified as low income, i.e. countries with a gross national product (GNP) per capita of US$785 or less in 1997 (World Bank, 1999). The actual GNP per capita of these African countries ranged from US$90 to US$750. The rest of the African countries had medium incomes, ranging from US$1050 to US$4230. The GNP per capita for Africa as a whole is US$500. Compare this income with US$970 for East Asia and the Pacific, US$390 for South Asia, US$3880 for Latin America and the Caribbean, US$25,700 for high-income countries and US$44,320 for Switzerland. In many African countries, a large proportion of the population live in severe poverty. The poorest countries include Guinea-Bissau where over 88% of the people were estimated in 1991 to live on less than US$1 per day, Zambia where the proportion was 85% in 1993, Madagascar where it was 72%, and Uganda where it was 69% in 1989–1990. Even in Kenya and Zimbabwe, which are generally regarded as being more developed than most African countries, 50 and 41% of the respective populations live on less than US$1 per day. In most of the African countries for which data are available, the vast majority of the people lived on less than US$2 a day, e.g. Ethiopia, 89%; Lesotho, 74%; Niger, 92%; Zambia, 98% (World Bank, 1999).

Gross domestic product
The same sorry picture emerges when we consider gross domestic product (GDP). The Netherlands with only 2.6% of Africa's population has a larger GDP than all the African countries put together (World Bank, 1999). Table 11.1 compares GDP per capita of a number of regions and countries. Africa's GDP per capita of US$522 is about one-tenth of what is considered by UNDP as the minimum income for a reasonable standard of living. Africa's GDP per capita is the second lowest in the world, after South Asia's US$373. It is only 4.8% that of Greece, and 1.6% that of Japan. Between 1980 and 1997, Africa's GDP grew at the rate of 1.7–2.1% annually, a rate well below the rate of population growth (2.9–2.7%).

Table 11.1. Gross domestic product per capita of
selected countries and regions.

Region/country	GDP per capita 1997 (US$)	Population (millions)
South Asia	375	1,289
Africa	522	614
East Asia and the Pacific	897	1,753
Latin America and the Caribbean	3,797	494
Greece	10,828	11
The Netherlands	22,530	16
Japan	33,346	126
Switzerland	41,914	7
World	4,830	5,829

Source: World Bank (1999).

Health

When we turn from the economy to health, we find that the situation in
Africa is equally depressing. Thus life expectancy at birth is only 53 years
in Africa, compared with 62 years in South Asia, 70 years in Latin
America and the Caribbean, and 78 years in high-income countries (World
Bank, 1999). Africa has the highest child mortality rate in the world. In
1996, the mortality rate for children under 5 years was 147 per 1000
live births. This contrasts sharply with seven per 1000 in high-income
countries. Many of the diseases responsible for death and ill health
are preventable and curable. Examples are malaria, diarrhoeal diseases
and respiratory tract infections. Hunger and malnutrition are major
problems in Africa. One out of three Africans does not have enough to
eat. Malnutrition is widespread in African countries; for children under
5 years of age it varied from 15 to 48% in 1990–1996 (World Bank,
1999).

Education

Although Africa has made considerable progress in the field of education
over the last 30 years, both the extent and the quality of education are
woefully inadequate. Nearly half of the African population is illiterate. In
1995, about 45% of the adults (54% of the females and 35% of the males)
were illiterate. In many African countries, 70–93% of the women and
50–79% of the men are illiterate (World Bank, 1999). Primary school
enrolments declined from 90 to 78% for boys and from 68 to 65% for
girls between 1980 and 1993 (World Bank, 1997). Although secondary
school enrolments improved slightly during this period, less than 25% of the
age group were enrolled. School drop-out rates are high, especially among
girls.

Human development index (HDI)

The human development index (HDI) is a useful measure of the socio-economic development of a country. Developed by UNDP, 'the HDI provides a snapshot of the level of human development in 174 countries ranked on a global scale' (UNDP, 1995). It combines the health status as indicated by life expectancy at birth, the extent of education as indicated by adult literacy and primary, secondary and tertiary enrolments, and the per capita income of a country. The goals set for development are: an average lifespan of 85 years, access to education for all and an income necessary for a decent standard of living. The average global real GDP per capita is considered as the minimum income for a reasonable standard of living. This is the purchasing power parity for an individual, which was estimated to be about US$5000 in 1992 (UNDP, 1995).

In 1995, the HDI varied from 0.185 in Sierra Leone to 0.960 in Canada (UNDP, 1998). Africa is the least developed continent: the average HDI value of 0.386 is below that for all low human development countries (HDI value 0.409). Of the 40 least developed countries in the world, in terms of HDI, 78% are African; and of the least developed 20 countries, 90% are African. Table 11.2 shows the HDI values of selected countries and regions.

Thus, Africa is exceedingly poor and underdeveloped whatever measure one uses – whether the more traditional measures such as GNP and life expectancy, or the HDI. The important questions are: why is Africa so underdeveloped and what are the prospects for development?

Table 11.2. Human development index values for selected countries and regions.

Country/region	HDI values 1995
Canada	0.960
Norway	0.943
Republic of Korea	0.894
Brazil	0.809
Zimbabwe	0.507
Nigeria	0.391
Côte d'Ivoire	0.368
Sierra Leone	0.185
Africa	0.386
Industrial countries	0.911
World	0.772

Source: UNDP (1998).

Why is Africa so poor?

If our work as scientists is to help improve the quality of life, we need to understand the socio-economic and political contexts in which it is done. If we are to contribute to the development of poor countries, such as Africa, we must understand the fundamental causes of poverty and under-development. The causes of poverty in Africa are as many as they are complex. The situation is so complex that it is often difficult to tell which is cause and which is effect. As an example, backwardness in science and technology is both a cause and an effect of underdevelopment. Africans often blame Africa's underdevelopment on the West, while the West blames it on the Africans. The truth seems to be that *both* the Africans and the West are to blame, as will become clear when we consider the main causes of poverty.

I believe that four factors are fundamental causes of poverty in Africa. They must be addressed as a matter of urgency if Africa is to overcome poverty and underdevelopment. These are bad governance, an unfair inter-national economic system, rapid population growth and backwardness in science and technology.

Bad governance

Bad governance is unquestionably the most important single cause of Africa's socio-economic wretchedness. Since its colonialization by western nations in the second half of the 19th century, Africa has been the victim of oppression and exploitation. Since independence was restored to Africa 30–40 years ago, most African economies have been ruined by repressive, autocratic, corrupt and incompetent governments. Autocratic rule has required huge military expenditures and it has caused political instability, displacement of millions of people, capital flight and demoralized popula-tions who are only concerned with survival.

Corrupt government leaders steal public resources intended for devel-opment and provision of public services. They collaborate with foreign multinational corporations and foreign agencies to siphon off billions of dollars from the African economy each year. One important way in which they do this is large so-called development projects that are not viable. Such projects are funded by loans from foreign governments, foreign banks or international agencies. The budgets include bribes or 'commissions' for government leaders, and huge salaries and expenses for the mandatory foreign experts. One study found that in 1988, technical assistance programmes in Tanzania spent at least two-thirds of the budget of US$300 million on the 1000 foreign staff, leaving only one-third or less for actual development work (Jolly, 1989). The total cost of the entire Tanzanian civil service was US$100 million. Whether projects succeed or fail – and a large proportion fail – the countries must repay the loans plus interest. One result is that development does not occur. The other is that Africa is saddled with a crippling burden of foreign debts. In 1995, these amounted to about US$226.5 billion which was 81.3% of Africa's GNP and 241.7% of its exports of goods and services (World Bank, 1997).

Who is responsible for bad governance in Africa? African leaders are to a large extent responsible. However, the West shares responsibility for bad governance. For political, strategic and economic reasons, western nations have played a key role in creating and then perpetuating oppressive and corrupt governments that have impoverished Africa. The classic example is the regime of Mobutu Sese Seko in Zaire (now the Democratic Republic of Congo), one of the richest countries in the world in terms of natural resources, but currently one of the poorest and least developed. It is well known that Mobutu's oppressive, exploitative, 30-year rule depended on the support of the USA, Belgium and France.

African officials, northern companies and northern governments are equal partners in large-scale corruption, or 'Grand Corruption' as one author (Moody-Stuart, 1997) calls it. To win government contracts, northern companies pay bribes to African public officials. The more important the official the larger the bribe, with top officials reportedly getting 20% or more of a project's budget. Northern governments not only allow bribery, they also encourage it by recognizing bribes for tax deduction. An encouraging step was taken on 17 December 1997 when 29 OECD member countries and five other countries signed an agreement to outlaw bribery of foreign officials by their companies (*Daily Nation*, 1997). We await implementation of the agreement and the extension of the agreement and subsequent legislation to cover political party bosses.

Unfair international economic system

The international economic system operates on the basis of unequal exchange between the North and the South. Based on the theory of comparative cost advantage, it was created by the colonial powers to obtain raw materials in exchange for manufactured goods (Olofin, 1997). The net result was the transfer of resources from African and other colonies to Europe. The patterns of production and trade as well as the dependence of African economies on western economies created by the colonial powers continue to impoverish Africa. The North continues to determine both the contents and the terms of exchange. The North uses its economic and political power to determine not only what goods and services are exchanged but also the prices Africa will receive for its exports to the North. The North also determines the prices Africa must pay for imports from the North.

Trade barriers and control of capital goods, technology and finance enable the North to control the international economic system in its favour. The result is net transfers of resources from the poor South to the rich North. Examples are net transfers to the North of US$163 billion in debt-related payments in the period 1984–1988 and US$83 billion in trade involving 18 non-oil commodities in 1981–1986 (The South Commission,

1990). In 1989, there was a net outflow of resources amounting to US$6 billion from Africa to the North (*Daily Nation*, 1990). The loss must be much greater now given the larger foreign debts among other avenues for resource transfer.

In short, the West is responsible for creating and perpetuating, in conjunction with African elites, an economic system, the net effect of which is an outflow of resources from the increasingly impoverished Africa to the North. Many western people, who generally want to help Africa, find this difficult to believe, but there is good evidence in support of it. An example is the UK's policy of making a 40% profit on its official 'aid' to poor nations (British Broadcasting Corporation, 1992). In general, rich donor countries gain more from 'aid' than poor recipient countries. Hancock (1991) discusses some important ways in which this happens.

Rapid population growth

Although reliable figures are not available, it is generally agreed that in Africa the population has grown much more rapidly than the economy since the end of World War II. The results are: increasing hunger as per capita food production diminishes; growing poverty as both GNP and GDP per capita decline; escalating unemployment; and disintegration of social services as demand stretches them. Moreover, rapidly growing demands have made it extremely difficult for African governments to devote sufficient resources to development or even to maintenance of developments already achieved.

According to the World Bank (1999), between 1980 and 1990, the average annual population growth rate was 2.9% whereas the GDP growth rate was 1.7%. In 1990–1997, growth rates for the population and GDP were 2.7 and 2.1%, respectively. Africa's growth rate of 2.7% is the highest of any region in the world. Compare Africa's rate of 2.7% for the period 1990–1997 with, for instance, Latin America and the Caribbean's 1.7%, East Asia and the Pacific's 1.3%, high-income countries' 0.7%, and 1.6% for all low- and middle-income countries. Interestingly, the rapid population increase over the last 50 years seems to have resulted from improved health, education and living conditions in Africa that cut infant and childhood mortality and increased fertility (Mlay, 1997).

Consensus seems to be growing that the way to reduce population growth rates to manageable levels is to lower fertility by improving the quality of life of the people. This requires better healthcare, universal primary education, particularly for girls, improved incomes and provision of social security. Promoting human development thus will not only reduce population growth, it will also equip people for productive roles in the development of their communities and countries. Many now recognize that this approach is more effective than the more direct population control methods attempted by the West in poor countries. It is also the ethical approach to a complex personal, family and community matter. Aggressive promotion of contraceptives often hurts the moral values and cultural sensitivities of people and leaves them wondering about the motives of contraceptive promoters.

Backwardness in science and technology
In the modern world, science and technology are absolutely essential both to socio-economic development and to the provision of basic services, such as healthcare, telecommunications and transport. No nation can develop or provide such services adequately without sufficient capacity for science and technology.

Both science and technology are in a very rudimentary state in Africa (Kinoti, 1997). This is indicated clearly by the minuscule expenditure on research and development (R&D). According to Barre (1998), in 1994 US$2.3 billion were spent on R&D in Africa compared with over US$178.1 billion in North America, US$131.5 billion in western Europe and US$87.3 billion in Japan and the newly industrialized countries of Asia (NICs). Africa's expenditure was about 0.5% of the world R&D expenditure. This and the Middle East's 0.4% are the lowest shares of the world R&D expenditure and contrast with 37.9% for North America, 28.0% for Western Europe and 18.6% for Japan and the NICs. Only 2.2% of GDP was spent on R&D in Africa in 1994. Compare this percentage with 22.2% for each of western Europe and North America, 14.2% for China and 11.4% for Japan and the NICs.

Africa's scientific output as measured by the number of publications was only 0.8% of the world total in 1995. Whereas Latin America, Asia, western Europe and North America either maintained or increased their share of scientific publications, between 1990 and 1995, Africa lost a large part of its share.

A similar picture emerges when we consider technological output in terms of patents. Africa's contribution is negligible as is the number of personnel engaged in R&D. As Adeboye (1998) points out, the data on R&D in Africa are very inadequate and not always reliable. However, even without hard data, it is obvious that science and technology are grossly underdeveloped in Africa.

Africa is almost completely dependent on imported technology. This raises two major problems. First, Africa is too poor to afford anything like the quantity and quality of technology it needs. Most of the world's technology is owned by northern companies who develop it for commercial not humanitarian purposes. Secondly, imported technology is an important contributor to the net outflow of resources from the continent. As the UNDP noted in its Human Development Report (1993), technical co-operation projects 'serve the priorities of the donors rather than build up national capacity'. Such projects are very costly to Africa. Around 75% of project budgets goes into salaries and other personal costs of the foreign experts who are mandatory for these projects. The annual salary and costs of one foreign expert may exceed the entire budget of a government department. These projects also frustrate national experts because they employ expatriates even in situations where national experts are unemployed or underemployed. The report gives Mali as an example: in 1990, donors employed 80 expatriate doctors and other health workers 'even though 100 qualified Malian doctors were unemployed'. Where national and foreign experts are engaged in the same projects, there are enormous disparities between their salaries, which demoralizes national experts. Also, sometimes foreign so-called experts are less qualified than national experts.

For more than two decades, the international community has recognized the need to assist poor countries to build an adequate capacity in science and technology, but little has actually been done to do so. Instead, key western nations have built up their own R&D capacities in vital areas such as tropical agriculture and tropical medicine in order to sell appropriate goods and services to poor countries and to provide employment for their own people. In the Lagos Plan of Action for the Economic Development of Africa, African governments recognized the vital role of science and technology in development (Organization of African Unity, 1981). However, since the plan was published in 1981, not much has been done to develop R&D capacity in Africa. Indeed, due to economic and political problems, there has been a sad decline in the work of universities and research institutes, which are the important R&D institutions in Africa. However, in spite of the difficulties, Africa has a slowly growing stock of highly qualified scientists and engineers. Given conducive economic, social and working conditions they could make significant contributions to the socio-economic development of the continent.

AGRICULTURE AND DEVELOPMENT IN AFRICA

Agriculture and future prospects

Although it is faced with an unprecedented economic and social crisis, Africa has the potential to reverse the present situation. The potential lies in 'its vast, poorly exploited resources of land, water, minerals, oil and gas; in its under-utilised people; in its traditions of solidarity; and in the international support it can count on', to quote a study by the World Bank (1989).

The study identified, correctly many will feel, agricultural production as the primary source of economic growth, at least in the short and medium term. An annual growth rate of at least 4% in agricultural production should enable African countries to meet their food needs and to generate foreign exchange to finance development. The proposed growth rates are realistic: 4% for agriculture, 5% rising to 7–8% for industry, and an overall 4–5% for the African economies.

Many will also agree with the strategy proposed by the study. The strategy calls for the creation of an enabling environment for both farmers and business people, and particularly the provision of physical infrastructure and incentives for private enterprise. Secondly, it calls for the building of the capacities of government institutions and the private sector, including private firms and non-governmental organizations (HGOs). Thirdly, the strategy requires that high priority be given to health and education, including science and technology education, in order to improve productivity. For this to happen, African countries must have competent, efficient and honest governments. Fourthly, agriculture must receive particular attention. Its growth requires an enabling environment of appropriate policies, credit, land tenure and rural infrastructure. It also requires new technologies, better trained researchers, farmers and extension agents as well as more effective farmers organizations. Attention must also be given to environmental protection because sustainable agriculture requires a sound environment.

The strategy calls further for enhancement of African entrepreneurship; exploitation of Africa's vast mineral wealth, including oil; expansion of energy supplies; economic integration of African countries to create larger markets within Africa; raising of the levels of investment, savings and foreign exchange earnings; raising of the level and focus of official aid; and concessional debt relief to ease the burden of debt service payments.

Contribution of agriculture to African economies

Agriculture is the mainstay of most African economies. It is the only, or the main, source of an income for about 70% of the population. Table 11.3 shows the contribution of agriculture to selected African economies. For Africa as a whole, the contribution of agriculture to GDP in 1995 was 20%. Compare this with 2% in high-income economies, 10% in Latin America and the Caribbean, 18% in East Asia and the Pacific, and 30% in South Asia (World Bank, 1997).

Contribution of livestock to African economies

Livestock production is the principal form of agriculture in several countries, contributing over 50% of the agricultural production. It was estimated in 1988 that livestock production contributed 8% of Africa's GDP and 25% of total agricultural production. If we take into account the contributions of draught power and manure to farming, the contribution of livestock to agricultural production comes to about 35% (Winrock International, 1992). The contribution of livestock to agricultural production in selected African countries is shown in Table 11.4.

Table 11.3. Contribution of agriculture to gross domestic product in selected African countries in 1995.

Economy	Contribution of agriculture (%)
Ethiopia	57
Uganda	50
Ghana	46
Malawi	42
Mozambique	33
Cameroon	39
Senegal	20
Kenya	29
Zimbabwe	15
Lesotho	10
Botswana	5
Africa	20

Source: World Bank (1997).

Table 11.4. Contribution of livestock to agricultural production in selected African countries in 1988.

Country	Livestock contribution (%)
Botswana	88
Lesotho	70
Sudan	58
Mali	44
Zambia	32
Tanzania	23
Uganda	14
Gabon	10
Liberia	9
Zaire (Congo Democratic Republic)	5
Africa	25

Source: Winrock International (1992).

Other contributions of livestock

In addition to providing food, high-quality nutrients, incomes, draught power and manure, livestock are important to Africa in other ways. First, they provide security against crop failure due to drought, pests or other problems. Therefore, in mixed crop–animal farming systems, it is difficult to persuade farmers to give up livestock farming even where crop farming is clearly more profitable.

Secondly, in many traditional societies, livestock has value which cannot be measured in economic or material terms. Thus dowry, which is still essential to marriage, could or can only be paid in livestock. Animals play a unique role in the social and religious life of society. Cleansing ceremonies and certain solemn vows require animal blood or animal victims. In addition, families love their animals, knowing each one by name and even talking to them as to a close friend. This non-material or 'spiritual' value of animals is on the decline as social and moral values are replaced by materialistic ones, to the impoverishment of the human spirit.

Enhancing the contribution of livestock to the African economy

A study found that there was excellent potential to increase livestock production and its contribution to African economies (Winrock International, 1992). It identified three areas of growth: expansion of mixed crop and livestock farming in the subhumid zone and the wetter parts of the semi-arid zone; increased productivity in the highland zone by greater use of technology and agricultural inputs; and expansion of intensive commercial production systems for poultry and pigs.

A growth rate of 4–5% a year in agricultural production, including livestock production, is an achievable target. For livestock, the target can be met if: (i) feed quality and supply are improved; (ii) animal diseases, particularly vector-borne diseases, are controlled; (iii) the essential institutions – research, extension and animal healthcare services – are strengthened; and (iv) an enabling environment of economic policy is created.

To these technical solutions, we must add political and economic solutions. First, Africa needs good governance if there is to be peace, political stability and the good management of public affairs that is essential to economic and social progress. Secondly, the international economic system must be reformed to make it fair to African and other poor countries. Trade barriers need to be lifted. The market forces need to be made kinder to the poor if Africa is not to be marginalized further. As Africa struggles to replace autocratic rule with democratic government, will multinational megacorporations become the new masters or will the people be able to take charge of their lives? Their ever growing power and influence – economic, political and social – must be brought under control.

SOME ISSUES

Attempts to increase animal production in Africa will create or intensify a number of issues. They include changes to African traditional values, environmental issues, ethical problems arising from manipulation of genetic materials, and the problem of poverty. Let us look briefly at each of these issues.

Social and moral values

As already stated, in many African traditional societies, domesticated animals are much more than an economic asset. They have unique social and 'spiritual' values. In some societies, animals are so intimately part of human life that they, like humans, must be protected against evil spirits, the evil eye and the consequences of human sin. This attitude ensures, or at least encourages, humane treatment of animals. A humane attitude is reinforced by the belief, probably based on the mysteries of mating and birth, that animals are made by a Creator to whom they ultimately belong.

However, traditional values are being replaced by materialistic values. The change is evident, for instance, among Kenyan smallholder farmers who practise zero-grazing for milk production and who view their animals from a strictly economic angle. A similar change, which also affects animal welfare, is seen in intensive poultry production methods. While it is hard to see how the traditional values can be maintained, it is evident that their erosion leads to the spiritual and moral impoverishment of African society and the detriment of animal welfare. New values that will enhance both human and animal welfare need to be developed. Materialistic values seem to be filling a moral vacuum created by the rapid decline of traditional values. Human experience, I believe, shows that a society's moral values are determined by its beliefs, whether they be religious or secular. The decline in African traditional values is evidently due to the erosion of African traditional religion.

As in all human cultures, neither African religion nor the moral values based on it were perfect. Some aspects were good and others bad. As an example of bad values, the Maasai people traditionally valued cattle above everything else; not even members of a man's family apparently mattered as much as his cattle. These people believed that *Enkai* (God) had given all the cattle in the world to the Maasai and so when their warriors raided other ethnic communities and carried away cattle, that was not theft; it was reclaiming their possession. However, it was wrong for a Maasai to steal from another Maasai and the culprit was required to repay ten animals for every animal stolen. Incidentally, few warriors could afford such heavy restitution, and their family or clan had to contribute, which kept stock theft very much under control in Maasai country.

Moral and ethical considerations are basic to social and economic development, as suggested earlier in this chapter and discussed more fully by Goulet in Chapter 7. It is clear that in every society and community, whether traditional or modern, there are both good and bad moral values. It is equally clear that people do not always put good values into practice. Even the scientific community is no exception, as we are reminded from time to time by fraudulent publications in scientific journals.

I believe everyone would agree that the rapidly changing world in which we live needs to find or recover correct moral values to govern individual lives, as well as interpersonal, economic and political relationships. We need moral values that are widely, if not universally, accepted. The question is what is to be the basis or foundation for such values. There is no agreement about this fundamental question. Some scientists, for example, have suggested that biological evolution could provide such a basis, with what promotes evolution being regarded as good and what hinders it as bad. Not everyone, however, can see why evolution is good. Indeed, many may feel that the evolutionary process, popularly known as 'the survival of the fittest', is itself evil. Some social scientists consider societal consensus to be the best or only basis for moral values. However, given the basically selfish nature of human beings, which Dawkins (1989) attributes to genes, such consensus is hard to reach. And consensus can sometimes pervert morality, as shown, for example, by Nazi Germany, slavery in the USA, apartheid in South Africa, genocide in Rwanda and communal violence in many parts of the world.

In most societies, religion traditionally has provided the foundation for morality, with God being seen as both the ultimate source of moral values and the final judge of moral conduct. This gave morality an authority no human being or institution could give. In Africa, this is still the situation, although traditional religion has been replaced by Christianity and Islam to a large extent. Western secular education has brought to Africa strengths, such as a better scientific basis for agriculture and a great broadening of the intellectual horizons, as well as weaknesses, such as a materialistic reductionism that sees a human being as nothing but an animal and that has no answer to the moral and ethical problems facing all humankind today. Religion has sometimes been abused; for example, Islam to support violence in the name of *jihad* ('holy war') or Christianity to support apartheid in South Africa. These lapses, however, do not invalidate religion any more than misuse of science or the occasional fraud in science invalidates science.

The question of morality requires urgent attention if the African people are to overcome poverty and enjoy genuine well-being. There are two reasons for this conviction. First, much of the poverty in the world today is due to inequitable distribution, which in turn is attributable to moral irresponsibility on the part of those who control the generation or distribution of resources. Also moral irresponsibility on the part of some individuals is frequently the reason for their poverty as well as that of their family. Secondly, it seems clear from the experience of the West, particularly since the end of World War II, that material prosperity alone is not enough. Individual as well as community well-being require not only material but also moral and spiritual well-being. History, I believe, shows clearly that a strong moral fabric is essential to a truly human life.

The traditional African moral fabric is decaying fast. So, what do we do? To an African Christian, the Bible is remarkable in that it articulates the same basic moral principles as were taught by our African ancestors. Like our ancestors, the Bible also identifies God as both the source of morality and the final judge of moral conduct. In numerous passages, the Bible insists on economic justice and on the moral responsibility to assist the poor and other disadvantaged and oppressed people. It seems that Christians from many other cultures have a similar experience with the Bible. This indicates that there are universal moral principles. We need to rediscover these moral principles in order to ensure that absolute poverty is eliminated from the whole of the world. Those principles, if sincerely followed, will also help us to reduce relative property and the glaring social equalities among and within nations. Africa's hope, I believe, lies in a rediscovery of the moral as well as the spiritual principles taught in the Bible (Kinoti, 1994). The West and other materially prosperous parts of the world also need to discover or rediscover these principles in order to avoid an empty, meaningless life. They need to discover the deeper meaning of life, for as Jesus Christ and the great teachers of humankind, such as Mahatma Gandhi, teach us, human life is much more than material possessions.

Environmental issues

There is some concern that increasing animal production will have serious environmental consequences, particularly in the rangelands and the sub-humid zone. While some scientists believe that livestock cause or can cause desertification of African rangelands, others argue that serious damage to the rangeland environment is caused not by livestock but by cultivation, tree-felling and other human activities. There are similar differences of opinion regarding the role of livestock in the destruction of tropical rain forests and the relationship between livestock and wildlife.

Animal trypanosomiasis illustrates the kind of environmental dilemmas that face African governments. The disease regulates livestock populations in, or excludes them from, a very large portion of African rangelands which are otherwise suitable for animal production. In the opinion of some scientists, this is a good thing for it protects the fragile environment. The question is whether to control the disease in order to increase animal production and in so doing risk environmental damage or whether to try to maintain the status quo in the face of increasing population pressure and poverty. Increasingly, poor families are migrating into tsetse-infested rangelands which they clear of bush for cultivation, thus eliminating tsetse flies. The environmental degradation consequent upon cultivation and other human activities is obvious.

Given the enormous population pressure in many African countries, human settlement in marginal lands is inevitable. Educating and assisting the people to employ productive and environmentally sound animal husbandry and other farming techniques would seem to be the best option. A major problem is that the governments do not have the financial resources.

In many African traditional societies, certain beliefs, taboos and attitudes helped to preserve nature. Thus, for generations, pastoralists and their livestock coexisted happily in the rangelands. For instance, in Kenya, only really poor families or ethnic groups lived by hunting; others did not eat wild animals. Incidentally, this custom seems to have protected people from some zoonoses, such as trichinosis. When, under the conditions of the Mau war of liberation in the 1950s, some youths on the slopes of Mount Kenya breached the custom and evidently ate a bush pig, they developed acute trichinosis, the first record of the disease in Africa south of the Sahara (Nelson, 1972). Attempts should be made to understand and preserve the good customs before they are lost from the rapidly changing cultural scene.

Genetic improvement of livestock productivity

Africa depends very largely on its indigenous breeds of cattle, goats and sheep for meat and milk production. While these breeds have considerable advantages over exotic breeds, particularly disease resistance, heat tolerance and ability to utilize low-quality feed, their productivity is low. Production can, and in the foreseeable future must be, increased mainly through disease control, improved feed and better management.

At the same time, genetic limitations to the productivity of indigenous breeds need to be addressed. Selective breeding can improve beef productivity of indigenous zebu cattle, as has been shown with the Boran breed in Kenya, but such breeds may need more resources and management than are available to the average African farmer. Cross-breeding of indigenous and exotic breeds increases productivity but it also tends to reduce disease resistance and other advantages of the native breeds. The challenge is to improve productivity without losing the important traits of the indigenous breeds.

As Heap and Spencer show so clearly in Chapter 2, biotechnology has great potential for increasing livestock productivity. It could make a very important contribution if, for instance, by nuclear transfer it increased the milk and meat productivity of African cattle. Similarly, biotechnology could increase the disease resistance and heat tolerance of exotic breeds such as the Ayrshire or Jersey, making it possible for pastoralists in East African rangelands to benefit from these productive dairy animals. In either case, introduction of genetically engineered bacteria producing highly efficient digestive enzymes into the digestive tracts of the animals could enable them to make better use of the generally low-quality feed available in those areas. It could also help to increase production by generating reliable diagnostic reagents and effective vaccines.

However, Africa is unlikely to benefit much from modern biotechnology, for a number of reasons. First, the technologies developed in the

West are bound to be too expensive for the majority of African farmers, who are very poor. Secondly, it will not be profitable for western biotechnology companies and research institutions to develop technologies for African farmers who cannot afford to pay for them. Western pharmaceutical companies, for example, do not find it attractive to develop drugs for such important tropical diseases as trypanosomiasis, human schistosomiasis and East Coast fever. Thirdly, even if technologies were available and affordable, many African countries do not have the technical capacity to adapt and apply them to their specific needs. If Africa is to benefit from biotechnology, the starting point should be to increase Africa's capacity to select, adapt and use the required technologies.

Finally, while acknowledging the potential benefits of biotechnology, I would like to express a concern. A fundamental problem with biotechnology is our inability to be certain where it will lead with regard, for instance, to environmental safety, but particularly with regard to ethics. Biotechnology now has the power to clone human beings. The dilemma about whether or not to clone humans points to the wider problem of whether it is necessary or feasible to set limits to what science can do. To put the concern differently and perhaps more popularly, should scientists 'play God'? The situation is especially worrying because modern science, despite its Christian origins, seems to be increasingly undermining the religious basis of morality without being able to offer a viable alternative.

Poverty

African nations and their friends need to ensure that any gains from increased animal production are directed towards the elimination of absolute poverty and the reduction of social and economic inequalities between the rich and the poor. UNDP (1997) quotes a Colombian educator as saying, 'Poverty is criminal because it does not allow people to be people. It is the cruelest denial of all of us human beings'. Poverty is a major moral issue because firstly, abject poverty dehumanizes people. It stops people from realizing their physical, intellectual, social, moral and spiritual potential. It does not allow them to enjoy a happy life, which includes good health, education, a decent standard of life, self-esteem and the esteem of others, or effective participation in political and other community decisions which affect their lives. Secondly, it is a moral issue because in many situations poverty is the result of inequitable distribution of resources which should benefit everyone. Inequitable distribution is often the result of exploitation of poor and powerless nations by rich and powerful nations. It is also often the result of poor people being exploited by their rich and powerful compatriots.

SUMMARY

Scientific work has real significance only if it contributes to human welfare and particularly to the improvement of quality of life for the poor and disadvantaged. Although this chapter is concerned primarily with the quality of economic and social life in Africa, it is important to bear in mind the fact that human life is a complex of many mutually interactive components, namely, the economic, social, political, cultural, moral, intellectual and spiritual aspects.

If our scientific work is to be relevant to development and the improvement of quality of life, we need to understand the economic, social and political contexts in which we work. A brief review of socio-economic conditions in Africa shows the continent to be in serious trouble. In many African countries, the majority of the people live in extreme poverty, i.e. they live on *less than US$1 per day*. This is reflected in a GNP per capita of US$500, which is less than 10% of the world average GNP per capita. A similarly depressing picture emerges from a consideration of GDP. For instance, The Netherlands, which has only 2.6% of Africa's population, has a larger GDP than all the African countries combined. Africa's GDP per capita is only 10% of the minimum income necessary for a decent standard of living.

Other socio-economic indicators confirm that, despite its vast natural resources, Africa is desperately poor and underdeveloped. These indicators include life expectancy at birth, infant mortality, literacy levels, school enrolments and the HDI, which combines health, education and income levels.

African leaders and foreign interests are principally to blame for Africa's plight. Four basic causes of poverty and underdevelopment in Africa are considered, namely, bad governance, an unfair international economic system, rapid population increase and backwardness in science and technology. Oppressive, incompetent and corrupt governments have hindered development, caused loss of resources to foreign interests and accumulated crippling external debt burdens for their countries. For economic, strategic and political reasons, western nations first created and then perpetuated oppressive and corrupt governments in Africa that continue to impoverish the continent. Similarly, the West is responsible for creating and then, in conjunction with African elites, running a world economic system of unequal exchange that impoverishes Africa and other poor countries. Thirdly, a population that is growing much faster than the economy has put unbearable pressure on Africa's resources, not only depleting resources but also retarding development. Fourthly, backwardness in science and technology severely constrains Africa's ability to provide basic services such as healthcare and telecommunications. Further, a lack of capacity for science and technology denies Africa the basic means of socio-economic development.

However, Africa can overcome poverty and underdevelopment. It has vast natural and human resources with which to do so. Agriculture is the mainstay of African economies and it is probably the only basis for economic growth in the short and medium term. An annual growth rate of 4–5% in agricultural production could meet food needs and generate foreign exchange to finance development, including industrial development. Growth in agricultural production calls for: an enabling environment of policy; credit to farmers; land tenure and adequate roads and other rural infrastructure; new technologies; better trained researchers, farmers and extension agents; and environmental protection for sustainable agriculture.

Animal agriculture makes a large contribution to agricultural production in Africa. In several countries, livestock production is the main form of agriculture. Overall livestock contributes 25% of Africa's agricultural production. In addition to their economic or material value, livestock play a unique role in the social and religious life of African society.

There is excellent potential to increase livestock production and its contribution to socio-economic development. An annual growth rate of 4–5% is possible if important diseases are controlled, feed quality and supply are improved, and research, animal healthcare and extension services are improved. Technology, especially biotechnology, could make major contributions to livestock production by genetic improvement of the productivity of indigenous breeds, and development of diagnostic reagents and vaccines. However, for technical solutions to be effective, Africa must have an enabling political and economic environment. This requires the combined efforts of African leaders and the international community.

Attempts to increase animal production will create or intensify a number of important issues. The increasing replacement of traditional values by materialistic values is bound to impoverish moral and spiritual life, and, indeed, life as a whole. These changes are also likely to affect animal welfare adversely. Population pressure and the decay of traditional beliefs and taboos may lead to serious environmental problems, particularly in rangelands. While biotechnology has great potential to increase animal productivity and thus production, Africa is unlikely to benefit much from it under the prevailing economic and technological conditions. The necessary technologies could at present only be developed by western companies or research institutions, and Africa has neither the resources to buy them in sufficient quantities nor the capacity to apply or adapt them to its particular needs. Finally, it is important to ensure that any economic gains from animal as well as crop agriculture are used to eliminate absolute poverty and to minimize inequalities between the rich and the poor.

Like the famous American economist Galbraith, we need to recognize that poverty is basically due to human nature, and that the fortunate, both individuals and nations, have a moral responsibility for the poor. In Africa and most other parts of the world, traditional moral values and the religious beliefs on which they were based are rapidly disintegrating. We need to discover spiritual and moral values that will enable our world to overcome the depressing problem of poverty. These values will also enable people to live a meaningful life in which the material as well as the spiritual, moral, social and intellectual needs are satisfied. The Bible articulates values which appear to be universal, and I believe therein lies hope for Africa and the other nations of the world. We need to discover those values and to put them into practice.

REFERENCES

Adeboye, T. (1998) Africa. In: *World Science Report.* UNESCO, Paris, pp. 166–180.

Barre, R. (1998) Indications of world science today. In: *World Science Report.* UNESCO, Paris, pp. 22–30.

British Broadcasting Corporation (1992) Panel discussion on World Service Radio, December 6.

Daily Nation newspaper (1997) Nairobi, 18 December, p. 12.

Daily Nation newspaper (1990) Nairobi, 14 July, pp. 10–11.

Dawkins, R. (1989) *The Selfish Gene.* Oxford University Press, Oxford.

Galbraith, J.K. (1998) On the continuing influence of affluence. In: *UNDP, World Development Report.* Oxford University Press, Oxford, p. 42.

George, S. (1986) *How the Other Half Dies: the Real Reasons for World Hunger.* Penguin Books, London.

Hancock, G. (1991) *Lords of Poverty.* Mandarin Paperbacks, London.

Jolly, R. (1989) A future for UN aid and technical assistance? *Development* 4, 21–26.

Kinoti, G. (1994) *Hope for Africa and What the Christian Can Do.* The African Institute for Scientific Research and Development (AISRED), Nairobi.

Kinoti, G. (1997) Africa must master science and technology to develop. In: Kinoti, G. and Kimuyu, P. (eds), *Vision for a Bright Africa.* International Fellowship of Evangelical Students Anglophone Africa and The African Institute for Scientific Research and Development, Kampala, pp. 153–192.

Mlay, W. (1997) The population question. In: Kinoti, G. and Kimuyu, P. (eds), *Vision for a Bright Africa,* International Fellowship of Evangelical Students Anglophone Africa and The African Institute for Scientific Research and Development, Kampala, pp. 126–152.

Moody-Stuart, G. (1997) *Grand Corruption.* Worldview Publishing, Oxford.

Nelson, G.S. (1972) Human behaviour in the transmission of parasitic diseases. In: Canning, E.U. and Wright, C.A. (eds), Supplement No. 1, *Behavioural Aspects of Parasite Transmission, Zoological Journal of the Linnean Society* 51, 109–122.

Olofin, S. (1997) The African economic crisis, debt burden and macroeconomic mismanagement by internal and external managers. In: Kinoti, G. and Kimuyu, P. (eds), *Vision for a Bright Africa.* International Fellowship of Evangelical Students Anglophone Africa and The African Institute for Scientific Research and Development, Kampala, pp. 63–103.

Organization of African Unity (1981) *Lagos Plan of Action for the Economic Development of Africa 1980–2000.* Institute for Labour Studies, Geneva.

The South Commission (1990) *The Challenge to the South.* Oxford University Press, Oxford.

UNDP (1993) *Human Development Report.* Oxford University Press, Oxford.

UNDP (1995) *Human Development Report.* Oxford University Press, Oxford.

UNDP (1997) *Human Development Report.* Oxford University Press, Oxford.

UNDP (1998) *Human Development Report.* Oxford University Press, Oxford.

Winrock International (1992) *An Assessment of Animal Agriculture in Sub-Saharan Africa.* Winrock International Institute for Agricultural Development, Morrilton, Arkansas.

World Bank (1989) *Sub-Saharan Africa: From Crisis to Sustainable Growth.* The World Bank, Washington, DC.

World Bank (1997) *World Development Report.* Oxford University Press, Oxford.

World Bank (1999) *World Development Report.* Oxford University Press, Oxford.

The Relationship of Ethics to Livestock and Quality of Life

12

E. David Cook

Whitefield Institute and Green College, Oxford University, Oxford, UK

INTRODUCTION

Science and technology continue to amaze and produce remarkable developments. Whether in medicine or in animal and food production, technology enables us not just to survive but to thrive. Inevitably, there is a downside to technology. It may produce new dangers and threats. This is exemplified in the possibilities of xenotransplantation – taking organs from animals to replace unhealthy organs in humans. Genetic manipulation of pigs can not only produce organs which will function normally in human beings, but can overcome many of the problems of rejection, which is the bugbear of all transplantation. Xenotransplantation is not new. For many years we have been using animal products in medicine and surgery ranging from insulin, to pancreatic islets and heart valves. However, the prospect of taking solid organs from animals to transplant into humans raises in sharp focus the moral issues relevant to livestock and quality of life in the 21st century.

Technology

We are able to do remarkable things. In food and plant production and animal breeding, the increasing application of genetic engineering is transforming agriculture and animal science. The moral question is whether there is a technological imperative. Just because we are able to do something technically, does not imply that we *must* do it. Immediately we are confronted with the need for some ethical framework or mechanism by which to decide whether or how far we should use or limit the use of

© CAB *International* 2000. *Livestock, Ethics and Quality of Life*
(eds J. Hodges and I.K. Han)

technology or biotechnology. The same kind of debate took place with the development of machinery in the Industrial Revolution. Now the debate is about our very survival as individual patients needing organs for transplantation or as producers and consumers needing food to live. The moral debate which has arisen from the technology of xenotransplantation may provide an entry point and focus for the wider issues of ethics, livestock and quality of life.

THE ETHICAL ISSUES IN XENOTRANSPLANTATION

In summary, it is possible to focus the debate about whether to use organs from animals in xenotransplantation into four broad categories (Nuffield Council on Bioethics, 1996).

Animal rights

This debate centres on whether there is a fundamental difference between humans and animals such that animals should not be used at all for food production, experimentation or xenotransplantation. The heart of the matter is not the degree of biological similarity or dissimilarity but the moral status of animals. It lies at the heart of the case for vegetarianism. It attempts to argue that animals have rights and humanity has responsibility to observe these rights.

There are few who argue that there is no difference between animals and humans. Those who do are often folk who feel closer to their pets than to human beings. Some argue that it is both dangerous and wrong to cross the lines between different species. Either there are appropriate natural differences or some religious point or purpose in the differentiation of species. Others accept difference, but argue on religious or cultural grounds that animals have a significance and should be preserved intact. None of these views is the majority one. That does not mean that minorities are necessarily wrong. It does pose proper questions about the fundamental differences between humans and animals i.e., whether we can make sense of the notion of 'rights' when applied to animals and if there is a conflict between animal and human need, which should take priority. The Kennedy Report on Xenotransplantation (Advisory Group on the Ethics of Xenotransplantation, 1996) denies that there are such things as animal rights, but makes the creative suggestion that animals do have the right to make a claim on humanity to be treated carefully and with respect. We do have responsibilities towards animals even if they have no rights in any meaningful sense. As with food production, there is a general agreement that in principle animals may be used for xenotransplantation purposes, if it is safe and if it will save human life.

Animal welfare

There is a degree of self-interest in keeping animals in good condition and good conditions. Healthy animals, kept in sanitary conditions, thrive. However, the reality of battery hen farming and the like and the occasional examples of outbreaks of disease keep the issue of animal welfare on the agenda. Animal welfare organizations argue beyond self-interest or public safety to human responsibility to care for the well-being of animals and to safeguard the conditions in which they are produced and maintained. In the UK, The Royal Society for the Prevention of Cruelty to Animals (RSPCA) believes that the welfare of farm animals must take into account five essential 'freedoms'. These were developed by an independent advisory body to the British government, the Farm Animal Welfare Council. The five freedoms are:

1. 'Freedom from hunger and thirst by ready access to fresh water and a diet to maintain full health and vigour.
2. Freedom from discomfort by providing an appropriate environment including shelter and a comfortable resting area.
3. Freedom from pain, injury or disease by prevention or rapid diagnosis and treatment.
4. Freedom from fear and distress by ensuring conditions and treatment which avoid mental suffering.
5. Freedom to express normal behaviour by providing sufficient space, proper facilities and company of the animals own kind'. (RSPCA, 1995)

In xenotransplantation, this means care about the conditions in which animals are bred, housed, cared for, operated on and killed. The stress is on using appropriate humane methods throughout. There will be debate about the extent and the application of such welfare freedoms for animals, but general agreement exists that animal welfare must be taken seriously not just for the benefit of the animals, but for the safety of humans. No one would countenance cruelty to animals, and there is an increasing awareness of the need to treat animals humanely and with respect.

Human safety

The threat of retroviruses is the main safety concern in xenotransplantation (Weiss, 1998). It is by no means clear what the effect or consequences will be of transplanting even a genetically modified animal organ from its natural medium and state into the human body. Current evidence shows that retroviruses are released when transplanting organs from one animal into a different animal. The fear is not just of harm to the recipient of the organ, but that what is triggered may be some new kind of virus, which is highly dangerous and easily transmissible. Once released, it might not be possible to control the spread of such viruses or retroviruses. The Ebola virus is one such frightening example. The main focus in the debate on xenotransplantation is not the moral status of animals or the morality of crossing species lines, but rather the practical questions of human safety. Food and livestock producers are increasingly aware of the need to take precautions to protect human beings. Outbreaks of bovine spongiform encephalopathy (BSE), new variant Creutzfeldt–Jakob disease (nvCJD) and the claimed links between diet and certain diseases creates a cautionary backdrop for livestock care and production. The problem is how far should reasonable caution go? In the ecological debate, there has grown up what is called the 'precautionary principle'. Crudely, it suggests that if there is a risk, then we ought not to develop that technology or biotechnology. If taken literally, this would end most scientific innovation and experimentation. That must not blind us to the distinctions between likely and possible risk, proven threat or unforeseen, and perhaps unforeseeable, happening.

Limiting harm and doing good

The fourth area of concern in xenotransplantation is the overwhelming need for organ transplantation. The sheer cost in human suffering of early death and long-term chronic illness points to the benefit of using every means to avoid such distress. There are various alternatives but these are either more costly or have severe drawbacks. The problem with transplantation is the scarcity of organs and donors. Xenotransplantation would alleviate the problem of the shortage of organs and remove the pressure of asking permission from grieving relatives to use the organs of a dead loved one. The motive for the use of animals is clearly to limit suffering and harm as well as to do good.

The fourfold analysis of the moral issues concerning xenotransplantation centre on animal rights, animal welfare, human safety and the moral principles of limiting harm and doing good. While xenotransplantation is very much a western issue, the need for and content of the moral debate around the area provides a point of entry into the wider issues of ethics, livestock and quality of life.

ETHICS

The philosopher usually begins by defining his or her terms. Other chapters have offered differing accounts of the nature and content of ethics. I wish to explore some of the concerns that confront us all in light of the problems facing livestock production and quality of life decisions in terms of animal production and care.

The need for ethics

As science has blossomed, the value systems of all nations are increasingly under stress. The impact of secularization worldwide has led to a shift away from traditional and religious perspectives on life, institutions and practices. In our global world, the main ethical forces seem to be individualism, consumerism, emotivism, utilitarianism and reductionism. It is immediately obvious that some of these are mutually contradictory but, in our so-called 'post-modern' world, even contradiction is embraced almost as a virtue.

Society stresses the importance of the individual. The driving forces of capitalism place the individual and his or her success and effort firmly at the very heart of morality. Right and wrong are based more on the freedom of individuals to create and choose their own ethical framework, rather than accept some form of natural law or more community-based traditional moral standards. Individualism leads to a consumerism, which is not so much interested in the staple requirements of life, but in quality of life, especially an ever increasing improvement in the standard of living and consumption patterns that accompany such a change. Morality is in danger of becoming purely subjective. It is construed as a matter of expressing our emotions and seeking to arouse the same emotions in other people. The language of good, bad, right and wrong no longer seems to rest on objective criteria, but rather to be a matter of taste, preference and expressing these preferences in words which are loaded with emotive power to make others react and feel what we feel and experience. There are serious implications for moral discussion and debate but, as the emotive perspective grows in power, societies increasingly seem to look to personal happiness and fulfil- ment as the full flowering of individual emotions. The problem is that if everyone seeks their own personal happiness and fulfilment, society polar- izes, collapses and disintegrates.

It is no accident, therefore, that utilitarianism, which pursues the greatest happiness of the greatest number, becomes an easy moral frame- work for a capitalist, consumer-based society. That means that animal production must serve the greatest happiness principle. It then runs the danger that the end justifies the means and any and every practice may be justified on the grounds of cost-effectiveness and efficiency. Yet there are strong reactions to the institutional power of utility and market forces. There is resistance to the reductionism which reduces morality simply to happiness or to any one feature. The complexity of human life and of moral reflection and practice mean that people are all too conscious of the need for holism, stressing the total nature of humanity in its whole context and variety. Discussions of the environmental problems facing humankind and the world, the loss of biodiversity, the need for wilderness, the status of animals, the need for justice and fairness all reveal that morality and ethics cannot be reduced to one aspect of experience, but must be dealt with in all their complexity.

Teaching and practising ethics

In teaching ethics to professionals ranging from doctors and nurses, to social workers, teachers, business people and agriculturists, it is vital to enable them to grasp the different approaches to ethics and the content of ethical reflection and practice.

At a very basic level, it is possible to describe three broad approaches to ethical reflection and practice. The first is to base our ethical decision-making and practice on fundamental principles. These are often remarkably similar in content. The medical profession no longer takes the Hippocratic Oath as such, but does seem to have general agreement over four key moral principles (Beauchamp and Childress, 1994). These are:

1. Do no harm – non-maleficence.
2. Do good – beneficence.
3. Justice.
4. Autonomy.

It is interesting to see how ethical reflection on xenotransplantation and on animal production might well take the principles of non-maleficence and beneficence in terms of animal welfare, human safety and using resources to limit harm and to do good. Inevitably, if we begin to ask how to apply these principles to systems of local, national or international production, we quickly have to offer some account of justice and of the limits (or otherwise) of autonomy or personal freedom, whether it is of the consumer or the producer.

The second approach to ethical decision-making stresses the importance of consequences in determining what is right or wrong, good or bad. What is right and good is what produces good consequences. Wrong and bad are the way we describe unpleasant consequences. Again it is easy to see how issues of animal welfare, human safety, limiting harm and fostering good can be understood very much in terms of the results and consequences of certain practices and behaviours. What is more difficult to square with a consequentialist approach is any full-blooded notion of justice. The fear is that we might engage in unjust activities or practices and justify that because the end result is in fact the greatest happiness of the greatest number. The problem is the vulnerability of the minority and the powerless. Ethical reflection which leaves no room or hope for such minorities and the vulnerable seems to be less than adequate or moral.

Disillusionment with the bankruptcy of moral discussion, debate and language either in terms of emotivism and utilitarianism or principles and consequences has led to a sea change in moral thought and practice. This change may have insight to offer in considering issues of livestock, animal production and quality of life. Alasdair MacIntyre's book, *After Virtue* (MacIntyre, 1985) suggests that morality is essentially about developing virtues or growing people. We might ask what is a good farmer or animal producer. We would then seek to describe the characteristics of a 'good' farmer. These would encapsulate the virtues of farming. They are the characteristics or virtues we ought to try to teach and encourage. When a person exemplifies and embodies these virtues, then we can see the kind of character a 'good' farmer is and has. If it is possible for us to reach some kind of agreement about the virtues, characteristics and character of a 'good' farmer or animal producer, we can then proceed to grow and develop such people. MacIntyre suggests that this requires a community which is bound together by a living, common story or narrative. We are not isolated units, but members of interlocking communities. Our personal story finds a place within a tradition and common story or history of a community of which we are part. That is not just where we belong and gain a sense of identity, but where we are grown and developed as moral beings.

This kind of ethical perspective offers animal producers a framework for discussion of how to define what is 'good', of the extent to which there is a community or a series of interconnected communities not just of producers and scientists but also of consumers. It also allows the exploration of a common, developing, dynamic story, which seems to be increasingly the case in light of the globalization of cultures, the power of the media, the common problems of ecological and environmental threat, scientific and technological developments, international trade and commerce and the common task of producing enough food for a hungry world.

Indeed, we might dare to go further and suggest that we have a common humanity, based both on the common needs of survival and flourishing and on common human experiences including moral, cultural and religious values. Part of our mistake is that we polarize the relationship between North and South, the developed and the developing worlds into a choice of (or being forced into the role of) dependency or independence. The middle way of interdependence might offer a better model for developing mutual and reciprocal practices, which did justice to our common humanity and our differing, localized expressions of our human experiences.

There seems an overwhelming case to teach ethics in the scientific and practical training of animal production. There is a crucial necessity for animal scientists and producers to behave in themselves and in their professional practice in the laboratory or on the farm and everywhere in between with high ethical standards. If this does not happen, then national and international regulatory authorities will impose controls on producers, which will be rigid, legalistically applied, inflexible and constraining. Unless animal producers put their own house in order and show clear, defensible moral practice and reflection, then they will be forced to accept unpalatable rules and regulation. Producer and consumer alike need a safety net of regulation, but good practice and ethical behaviour go beyond the merely legal and permissible, to what is genuinely in the common good in limiting harm and doing positive good in a just and fair manner.

THE COMMON GOOD

The core and content of the common good must be related to human flourishing and avoiding harm to humans, animals and the environment. We have an ethical responsibility for our world in terms of feeding the hungry, respecting the environment, animal welfare and preventing harm by ensuring that food and animal production are safe and secure, creating not just personal, company or commercial profit, but maintaining sustainability and biodiversity. These will require some view of stewardship, stakeholding, responsibility and respect. Part of this will be to consider issues of quality of life.

Quality of life

In health, welfare and development, there is much work being done on how to define and measure not just life itself, but the quality of life (Williams, 1995). In a sense, it is trying to answer the question, 'What is a good life?'. Like poverty, it is possible to interpret this in both absolute and relative terms. There are some things we absolutely need in order to survive, such as food, shelter and clothing. There are some things that we need and want in order to participate fully in the life of our particular society or community. We want what those around us normally have. We also want to have all the good things of life to enhance the quality of our lives. There is a distinction between what we need and what we want. It is unlikely that all of these aims can be fulfilled and achieved. However, we do need to explore what are the basic necessities for existence and to have a reasonable life. We then need to reflect how we can achieve that for all humanity. Then we can begin to consider what is an appropriate quality of life in our particular setting and context. There will and must be variations, but some of us may be more content with that variety than others. Part of the ethics of development is to consider not just how to develop the so-called 'underdeveloped' nations, economies and animal production, but whether these should be developed at all. James Baldwin, the famous black novelist, in his *The Fire Next Time* (Baldwin, 1962) asks the haunting question of whether black people in the USA really want to be integrated into a white society, which is not just being destroyed, but actually destroying itself. He asks if we want to be integrated into a burning house. It is by no means clear that the quality of life of the 'developed nations' is higher than that of the 'underdeveloped' nations in anything except material goods.

CONCLUSION

In the end, we are driven back to our ethical frameworks. We are part of local communities as well as of the common community of humanity.

We are responsible not just for ourselves, but we are also our sisters' and brothers' keepers. We ought not so much to demand our rights, but seek to fulfil our responsibilities and to respect our world and all that is in it, both animal and human.

SUMMARY

Using xenotransplantation as an exemplar, this chapter explores the moral issues at stake and seeks to apply them to ethics, livestock and quality of life. The need for ethics is explored and the nature and content of current ethical reflection described. This is then applied to a discussion of what is a 'good' farmer or animal producer. This leads to an analysis of the common good and of quality of life, arguing for sustained ethical reflection on commonality rather than difference. Responsibility and respect rather than rights are proposed as a way forward.

REFERENCES

Advisory Group on the Ethics of Xenotransplantation (1996) *Animal Tissue into Humans*. Department Of Health, HMSO, Norwich.

Baldwin, J. (1962) *The Fire Next Time*. Dell, New York.

Beauchamp, T.L. and Childress, J.F. (1994) *Principles of Biomedical Ethics*, 4th Edn. Oxford University Press, Oxford.

MacIntyre, A. (1985) *After Virtue: a Study in Moral Theory*, 2nd Edn. Duckworth, London.

Nuffield Council on Bioethics (1996) *Animal to Human Transplants: the Ethics of Xenotransplantation*. Nuffield Council, London.

RSPCA (1995) *Farm Animal Welfare: You Can Make a Difference*. RSPCA Information, P41 6.95, Horsham, West Sussex.

Weiss, R.A. (1998) Transgenic pigs and virus adaptation. *Nature* 391, 327–328.

Williams, A. (1995) *The Role of the EuroQol Instruments in Qaly Calculations*. Centre for Health Economics, University of York, UK.

Community of Life – the Ethical Way Forward

John Hodges

Mittersill, Austria

LEADERSHIP NEEDS AN ETHIC

Who are the stakeholders?

Successful solutions to the issues of livestock, ethics and quality of life raised by the authors in this book require new styles of leadership in scientific, social, economic, political and business areas. What types of leaders are needed? The editors and authors conclude that ethical leadership is essential. Ethical leaders seek the 'Best' decisions for the primary parties involved in any decision, and are also concerned with the impact and interests of all stakeholders. This is a truly ethical approach and requires defining stakeholders more precisely. We shall now seek to do this.

Specialization and reductionism in the 20th century has narrowed the definition of stakeholders to include only those with an active and immediate role or interest in any decision. Thus stakeholders in a business are the shareholders, executives, management and workers plus the banks which lend capital and any other bodies with a direct financial interest. Progressive businesses will seek to include customers among their stakeholders. However, business generally does not extend the list of stakeholders beyond those primary boundaries unless required to do so by law. For example, companies in the late 20th century have been required by legislation to consider the environment – and by extension the general population – as stakeholders. Thus, companies have been legally required to process effluents before discharging them into rivers, the air or coastal waters. Coal mines have been ordered to dispose of slag heaps of waste. Open-cast mining has to restore the land. Legislation on pharmaceutical drugs has required business to consider future generations as stakeholders even though they are neither born nor conceived and have no financial association with the business.

Ethical leadership sees the big picture and comprehends the overall sense in which the whole of humanity are stakeholders in the new scientific and economic developments. Even though biotechnology has come from western science laboratories, its impact – for good or ill – will be upon the whole world, for science knows no barriers. Early entry into market economy capitalism (MEC) has given the West huge accumulated capital and wealth which is a great advantage in launching economic globalization of food production and trade. Ethical leaders recognize that if this globalization is to be successful, it cannot be limited to the economic and financial interests of the primary stakeholders in the companies concerned. Ethical behaviour, as well as the economics and logistics of food production, have to be globalized. Ethical leaders recognize that globalization of biotechnology and of business will not work with the old boundaries of primary stakeholders. That limited view dominated the development of MEC at the national level until companies were required to broaden it by legislation. True globalization means recognizing that everyone has a stake in the outcome. Globalization which sees the shareholders of multinational corporations as the only stakeholders is bound to fail. Ethical leadership understands that decisions made by the powerful and wealthy will affect the weak and poor who are, in fact, the majority of stakeholders in the future of the world.

Awareness of world community

The term 'globalization' recognizes that biotechnology and business are no longer national but international activities. It should also mean recognition that we belong not only to a national community but also to the world. It is neither logical nor reasonable nor will it be successful to seek globalization of activities while limiting the concept of stakeholders to western nations because they were the wealthy investors with finance and technology. Ethical leadership means making decisions in the best interests of all the stakeholders.

Fear of the lack of such ethical leadership causes poorer developing countries to anticipate that biotechnology and pharmaceutical products may be sold to their citizens without adequate testing or even after testing has shown some risk. An example is the current massive promotion and sale of tobacco products in developing countries by multinational companies when the same companies are already required by law to warn citizens in their own countries of the dangers to health, and to pay compensation for premature death. Such a practice is unethical leadership. A similar fear is associated with the food products of biotechnology and with abuse of the environment. The sense of identity which has been engendered by belonging to a nation state must be extended to the world community. Otherwise globalization has little meaning more than economic exploitation and colonization.

Community is a difficult concept for individuals in western society. Both science and MEC have encouraged individual success and competition. Awareness of community does not mean relinquishing initiative and creativity. It means investing and effectively using individual skills, experience, knowledge, education, wealth and other resources to create new wealth in ways which will benefit the whole community and not only one group of those affected. Ethical leadership is based upon an awareness of belonging to community. Ethical decision making means thinking about the interests of all the stakeholders. Community is not a matter of idealism; it is the basis of human survival. The concepts and practice of community and of ethical leadership are extremely important. These concepts are so alien to western society as we enter the 21st century, yet so basic to the ideas of globalization, that this concluding chapter seeks to explore the importance of human community through a consideration of our relationship with domestic livestock.

DOMESTIC ANIMALS AND HUMAN PROGRESS

The history of civilization is closely associated with domestic animals. In the early days of human communities, around 10,000–12,000 years ago, a few large mammalian and bird species were domesticated, enabling humanity steadily to rise from primitive conditions to life of higher quality. Large domestic animals made possible the move from hunting, gathering and shifting cultivation to more settled lifestyles.

How have domestic animals played such a key part in the development of human community? What are the special contributions of animals to human well-being? Animals release people from the hard labour of heavy field work; animals make possible the transport of natural resources and farm products to other communities for barter or sale; animals provide animal fat and protein for improved nutrition; animal milk enables infants to survive and grow when quantities of human milk are insufficient; animals provide leather, wool and horn for clothing and shelter; animal fat is used for lighting; dried manure from large animals is fuel for cooking and heating; animal power is used for extracting water from the ground and from rivers for domestic use and for irrigation; animals contribute to improved and integrated farming systems on cropped land; ruminant animals harvest natural vegetation that would otherwise not enter the human food chain; throughout human history, riding animals was the fastest way to travel over land until the invention of the railway in 1829 – only 170 years ago. The domestication of animals was the first step to improving the quality of life through science and technology.

Today the majority of people in the world still depend upon animals for these services and, without them life, even in the simplest societies, would disintegrate again into the slavery of food production. The major advances in European civilization leading to trade, industrialization, the application of science and the development of MEC were possible because animals had first freed a proportion of the population from the daily routine of food production. Following further applications of science and technology throughout Europe and North America over the last 150 years, the majority of people have been set free from work on the land, leaving only 5–10% to farm. This fact can be traced back to the first step of domesticating animals. Freed from the necessity for each family to produce its own food, advanced societies have become immensely creative and modern life has become utterly different. Today, one has only to visit rural areas of Africa, Asia and Latin America to see the contrast with the West and the significant contribution of domestic animals. Closer to home, east of the heart of Europe, in many of the 14 new states of the former Soviet Union where the infra-structures of society have collapsed, one can also see the vital role of domestic animals permitting rural people to survive and to maintain human dignity in the current conditions of great poverty.

Influence of domestic animals on human values

For thousands of years, everyone was in touch daily with domestic animals. Since animals are a resource of such great value, it is easy to understand why people have held them in high esteem and have sometimes regarded them as sacred. People live in close contact with their animals. Usually each family has a few. Owners give animals food and care to ensure their health, longevity, and their ability to serve and to reproduce. Their value is recog-nized at special celebrations including birth, marriage and death. Animals are wealth, and are used both for savings and as currency. The status of a family or community leader is often recorded by numbers of livestock owned. In some parts of Africa today, a bride is given in return for livestock. In India, Hinduism, the major national religion, holds the cow in special honour and sees a link between the life of domestic cattle and human life. In Moslem society, sheep and goats are vital for religious obligations. In early Jewish periods, before AD 70 when Titus destroyed the Temple in Jerusalem, animal sacrifices were a central part of individual and commu-nity worship. Domestic animals have greatly influenced community rituals and values in most early societies.

We need to distinguish traditions and rituals of life from values. Tradi-tions and rituals are often beautiful and they mark for us the pattern of life, but they are rarely essential and we have dropped many of them from life in the West. In contrast, community and public values in a society, as the name implies, are extremely important. Every society has values. Values determine the direction which a society takes. Values enable a society to survive and advance – or they cause its decline. Values lie at the heart of a society and determine the goals people will work to achieve. Values direct activities. Values allocate resources in a society and thereby shape its nature.

The values of western society today are vastly different from those of Europe centuries ago. Values today are focused upon material prosperity, upon economic growth, upon gross national product (GNP) and upon the rights of the individual to do what he or she prefers with the rewards of labour and investment. In a democracy, society's values shape government policy and legislation. Many thoughtful people today are deeply concerned that our current, narrowly focused values in the West do not provide sufficient care for the environment and for animals and, further, that they define quality of life solely in material terms for immediate consumption.

What has this to do with domestic animals? In my view, the historic interrelationship of animals and people deeply influenced the way in which people see life – giving society its world view. The new western values and world view have come about partly because of the lack of daily contact with animals and the natural environment. In rural society, domestic animals provide the most personal and intimate connection people have with nature, due partly to the fact that humans and animals live and work together in daily contact. The fact that a person owns individual animals leads to a personal commitment to care for them. When people accompany cattle, sheep or goats into the natural environment for grazing, they realize that animals and human communities are parts of the whole natural order. People without animals are lost in slavery. Domestic animals need society for protection. Neither can live in a broken environment. Excessive use of one component, for example overgrazing leading to depleted vegetation, places human life and animals at risk. We are enough like animals to be kept humble; we are different enough from animals to be aware of our unique responsibility as 'husbandman' of the natural world.

Thus, the values of simpler societies for thousands of years were based upon a holistic view of life. Community embraced all individuals and everyone knew that each component of life is integrated and that life functions as a whole, like an organism with interdependent parts which must be sustained for life to continue. In the West, we have lost this world view. We discovered that by focusing upon one component we can make it more productive, but in our enthusiasm we forget the balance of the whole. It is the danger of reductionism. In earlier societies, the intimate dependence upon domestic animals gave more appreciation of the whole environment and helped society to realize that life is entwined with all the natural resources of the world. Although it is not so self-evident, the West today is still dependent upon natural resources. One cannot take endless quantities of everything without upsetting the balance and eventually precipitating a collapse that will reduce quality of life. The earth is in dynamic equilibrium. In tribes owning large herds of cattle, sheep or goats, the dilemma and tension are well known. The attractions of larger numbers of animals to ensure that some survive periods of drought have to be balanced against overgrazing and poorer quality animals. Those who prefer more and more animals nearly always lose. We, in the West, need to ponder the deeper implications of the lost relationship of western civilization with the environment, with domestic animals, with each other in our communities and with other societies on earth. I believe better understanding of these relationships is a key to our future options. Thinking more about where we have come from will help us to draw up a balance sheet of what the West has gained, what we are in danger of losing, where we are going and what goals we are advising poorer countries to pursue.

WESTERN SOCIETY TODAY

In the very, very recent past in the West, we have built a new type of urban lifestyle from which we physically exclude animals. In these new societies, we limit our contact with animals to buying animal products in shops and supermarkets where there is virtually no evidence of the animal itself. The proportion of the European Union population now involved in food production has been reduced overall to less than 10% and, with the option of importing food, sometimes only 3% of the national population produces food.

This revolution has been brought about by the replacement of animals in modern western society. We use fossil and nuclear fuels and hydro-electricity instead of animal power for farm work, general transport, cooking, light and heat; clothing of tropical plant or synthetic origin instead of local animal products; inorganic fertilizers on farms instead of animal manure; animal protein but not animal fat. International trade and refrigeration of food have released us from the need to keep animals nearby. We now want animals only for food and, as a by-product, for leather and wool. The increasing demand for sport and companion animals does not replace the older association as we bring these pets into our highly advanced isolation.

The West has experienced a revolution in the use of domestic animals. As recently as 200 years ago, everyone in town and country was in personal and daily contact with domestic animals. During my childhood in the UK, horses were working every day on all the roads in town and country. Today, people in western civilization live in comfort unprecedented historically, without contact with domestic animals. Animals have disappeared from the streets, from the town dairies which used to provide milk to urban populations, families no longer keep a pig or chickens, local abattoirs have been closed. On a trip through the country, the tourist sees only cows and sheep in the distance, and many of these are hidden in confinement along with pigs and poultry.

In the West now, we see animals solely as an economic resource, hidden away in remote locations with the objective of serving the market economy with animal products of high hygienic and eating quality. The market is uninterested in the other 'old-fashioned' contributions of animals to human life. Our isolation from the environment that supports us is evident when many urban dwellers are timid in the presence of domestic animals and when manure, animal smells and animal noises seem strange and embarrassing. Yet it is our isolation today that produces a truly negative result when tens of thousands of animals in one unit pollute the environment and harm health.

Contrasts

Although we may feel that our western society is more advanced and economically superior to people in less developed societies, in truth we are the sector of humanity that is now isolated and insulated from many of the real issues which lie at the heart of human survival. Our urban lifestyle is out of touch with the natural cycles of the seasons and the environment of the earth. Yet we need the natural resources of the world as much as ever for long-term survival. We live in a sort of plastic bubble where everything in life is controlled and the rhythms and uncertainties of nature have largely been tamed. People in developed countries comprise about 20% of the world population and consume 80% of world production of milk and dairy products and 66% of world beef production. These figures for animal products indicate our levels of consumption of most natural resources including energy, food and water. If we continue to inflate the bubble with gross overconsumption, it will burst. Then we shall lose our high-cost, isolated materialistic life and we shall find the natural environment deeply and permanently damaged – unsustainable. People who venture outside the bubble, physically or mentally, can already see this. The majority living only inside often do not want to know how dependent they are upon the continuance of a healthy environment outside. The earth systems are robust and can absorb much change; but massive overuse, pollution and abuse on an ever increasing scale can only lead to long-term breakdown that is irreversible on our time scale.

The application of science through MEC has changed farming from a way of life into a business for food and fibre production. The model of intensification and specialist large-scale animal production is being applied to the whole of life, and currently enables Europe, and the West as a whole, to enjoy a much higher quality of material prosperity than the majority of other people in the world. In the West, society's values and political policy now define the good life as increasing GNP and material prosperity. We have lost touch with the lessons of living with animals; namely that quality of life is not a solitary experience but flows into and from interdependence and community; and that sustainability must be based upon cycles involving preparation, harvest and post-harvest recovery. We cannot expect to milk a cow every day of her life from birth to death. The nature of things on this planet Earth demands that we invest not only hard work and skills but also time, patience and some restraint to gain the reward. The penetrating statement of Jesus 2000 years ago is still as relevant: 'Man cannot live by bread alone'.

Change in values in Europe

I will briefly review how rapidly values in Europe have changed. As students and young professional animal scientists after World War II, many of my colleagues and I were challenged by the shortage of food. Food rationing in the UK did not end until 1954, 9 years after the end of hostilities. We served the farmer, helping him use better science and technology. Through him, we served the consumer with cheaper food. The practice of farming was called 'husbandry'. The farmer was a 'husbandman', who cared for his animals, crops, land and water resources. He was the steward of the natural environment and resources temporarily in his care, using sustainable production methods. A good farmer produced good quality simple but varied and wholesome food, conserved the natural resources, with his family lived a satisfying life in rural surroundings and left his farm in better shape both physically and economically for his children.

Since the 1960s, the change in values has accelerated. Science has succeeded. Europe and North America produce too much food for home consumption. All focus is upon reducing unit costs. The farmer is now a business person 'husbanding' his or her financial resources. Universities have renamed Animal Husbandry as Animal Science. Science now serves business and business serves the market. Consumers, influenced by advertising, are kings and queens in the West today.

Of course it is wonderful that there are no longer hungry people in the West where we are light years away from hunger, while there are nearly one billion people suffering from undernourishment or starvation. The choices of animal products in our supermarkets are lavish beyond discretion and discernment. They are beyond the imagination of our grandmothers. More importantly, they are beyond the reach of 80% of the world who do not live in the West. The gap is increasing each year.

We pay a price. Actually, it will largely be paid by future generations, which makes it less threatening to us. Today our focus is upon profit, reduced unit costs, the desire to make more money this year than last and the supremacy of capital and its cost. All these pressures urge upon the livestock producer the need to succeed as a business person and to neglect the long-standing practices of sustainable production, care of animals and good husbandry of the environment. New scientific techniques and production methods for animals now regularly challenge the sustainability of the environment, put the animal in unnatural conditions, direct its hormonal system into new patterns, modify its genetic constitution and view the animal solely as a resource to be exploited for immediate profit and lower prices. Slowly, under pressure and with reluctance, some governments are legislating minimally against the most extreme practices in the interests of animal welfare and of the environment. Most politicians resist such legislation because of opposition from business and votes from a society largely ignorant and uninterested in these issues.

A life agenda driven only by values that maximize the material prosperity of the individual is a reductionist view. It takes no account of the larger whole. Yet, the history of human civilization is the story of community slowly built up by hard work and wisdom but periodically destroyed by narrow agendas and foolishness. Living as individuals alone in nature was a dangerous and precarious lifestyle. Civilizations progressed when quality of human life was defined to include transcendence as well as material prosperity. Civilizations declined when a material agenda and individual greed squeezed out higher values. Europe has a heritage which upgraded society over many centuries and defined quality of life in multiple dimensions. However, to our loss, we are neglecting our heritage and increasingly have tunnel vision for immediate and personal material prosperity.

We have lost touch with the values that our ancestors learned from their animals. They knew that if you want your cow to have a calf and to produce milk next year, you cannot take all the resources of the cow this year. Resources need husbanding if they are to produce sustainably in perpetuity. Natural capital can be squandered.

COMMUNITY OF LIFE AND ETHICAL LEADERSHIP

Under the influence of science and market economy pressures, values in Western society have lost the holistic approach. It seems totally irrelevant to the shopper buying animal products to suggest that this way of life is harming the environment. Like all societies, we are driven by our values, which are leading us from legitimate self-interest to greed. Greed always destroys and produces inequity. It is time to look back to our history and give more respect to the lessons from animals that were key guides to our ancestors on the meaning of wholeness, sustainability and community.

I am not advocating a return to animal power and primitive lifestyles. The world needs good science and responsible business to create wealth to raise the quality of life throughout our exploding world population. Rather, I am calling for higher values in western society in the business of creating and using new scientific knowledge, the new wealth. Better values are characterized by community, which means sharing and interdependence; by genuine self-interest in quality of life for all people instead of individual greed; and by patiently working with nature in the interests of sustainable use.

We are foolish to think that wholeness and sustainability are negative restrictions on the good life. Wholeness and sustainability are quality of life experiences not provided by the search for endless and greater material prosperity. Our current values in western society are divorced from real and long-term meaning and, pursued in singleness of mind, are unable alone to deliver quality of life. Our present route is not only inequitable; it lacks quality, and it is unsustainable. We must change – or unpleasant change will be forced upon us and our children.

We can learn much from the animals, as did our ancestors. The greatest lesson is that we are all part of a greater whole. As shown in this book, animal biotechnology and its emerging techniques, the intensification of animal production, and the globalization of food production and trade can bring great benefits to mankind provided that we see them as potential contributors to building community and not means of enriching one part of human society at the expense of another. Decision-making and actions with consideration for *all* the other stakeholders in community is the basis of ethical leadership.

ACKNOWLEDGEMENTS

Parts of this chapter are taken from a paper presented by the author at the International Congress on 'Regulation of Animal Production In Europe' held by Kuratorium für Technik und Bauwesen in der Landswirtschaft (KTBL) in Wiesbaden, Germany from 9 to 12 May 1999.

Index